PROGRESS IN CLINICAL AND BIOLOGICAL RESEARCH

1983 TITLES

Vol 115: **Prevention of Hereditary Large Bowel Cancer,** John R.F. Ingall, Anthony J. Mastromarino, *Editors*

Vol 116: **New Concepts in Thyroid Disease,** Roberto J. Soto, Gerardo Sartorio, Ismael de Forteza, *Editors*

Vol 117: **Reproductive Toxicology,** Donald R. Mattison, *Editor*

Vol 118: **Growth and Trophic Factors,** J. Regino Perez-Polo, Jean de Vellis, Bernard Haber, *Editors*

Vol 119: **Oncogenes and Retroviruses: Evaluation of Basic Findings and Clinical Potential,** Timothy E. O'Connor, Frank J. Rauscher, Jr., *Editors*

Vol 120: **Advances in Cancer Control: Research and Development,** Paul F. Engstrom, Paul N. Anderson, Lee E. Mortenson, *Editors*

Vol 121: **Progress in Cancer Control III: A Regional Approach,** Curtis Mettlin, Gerald P. Murphy, *Editors*

Vol 122: **Advances in Blood Substitute Research,** Robert B. Bolin, Robert P. Geyer, George J. Nemo, *Editors*

Vol 123: **California Serogroup Viruses,** Charles H. Calisher, Wayne H. Thompson, *Editors*

Vol 124: **Epilepsy: An Update on Research and Therapy,** Giuseppe Nisticò, Raoul Di Perri, H. Meinardi, *Editors*

Vol 125: **Sulfur Amino Acids: Biochemical and Clinical Aspects,** Kinya Kuriyama, Ryan J. Huxtable, Heitaroh Iwata, *Editors*

Vol 126: **Membrane Biophysics II: Physical Methods in the Study of Epithelia,** Mumtaz A. Dinno, Arthur B. Callahan, Thomas C. Rozzell, *Editors*

Vol 127: **Orphan Drugs and Orphan Diseases: Clinical Realities and Public Policy,** George J. Brewer, *Editor*

Vol 128: **Research Ethics,** Kåre Berg, Knut Erik Tranøy, *Editors*

Vol 129: **Zinc Deficiency in Human Subjects,** Ananda S. Prasad, Ayhan O. Cavdar, George J. Brewer, Peter J. Aggett, *Editors*

Vol 130: **Progress in Cancer Control IV: Research in the Cancer Center,** Curtis Mettlin, Gerald P. Murphy, *Editors*

Vol 131: **Ethopharmacology: Primate Models of Neuropsychiatric Disorders,** Klaus A. Miczek, *Editor*

Vol 132: **13th International Cancer Congress,** Edwin A. Mirand, William B. Hutchinson, Enrico Mihich, *Editors.* Published in 5 Volumes: Part A: **Current Perspectives in Cancer.** Part B: **Biology of Cancer (1).** Part C: **Biology of Cancer (2).** Part D: **Research and Treatment.** Part E: **Cancer Management**

Vol 133: **Non-HLA Antigens in Health, Aging, and Malignancy,** Elias Cohen, Dharam P. Singal, *Editors*

Vol 134: **Globin Gene Expression and Hematopoietic Differentiation,** George Stamatoyannopoulos, Arthur W. Nienhuis, *Editors*

Vol 135: **Developmental Pharmacology,** Stuart M. MacLeod, Allan B. Okey, Stephen P. Spielberg, Michael Spino, *Editors*

Vol 136: **Hypoxia, Exercise, and Altitude: Proceedings of the Third Banff International Hypoxia Symposium,** John R. Sutton, Charles S. Houston, Norman L. Jones, *Editors*

See pages following the index for previous titles in this series.

CHEMICAL TAXONOMY, MOLECULAR BIOLOGY, AND FUNCTION OF PLANT LECTINS

CHEMICAL TAXONOMY, MOLECULAR BIOLOGY, AND FUNCTION OF PLANT LECTINS

Proceedings of a Symposium
Sponsored
by E•Y Laboratories
May 31—June 3, 1983
Asilomar, California

Editors

IRWIN J. GOLDSTEIN
University of Michigan
Ann Arbor

MARILYNN E. ETZLER
University of California
Davis

ALAN R. LISS, INC., NEW YORK

Address all Inquiries to the Publisher
Alan R. Liss, Inc., 150 Fifth Avenue, New York, NY 10011

Printed in the United States of America.

Library of Congress Cataloging in Publication Data

Main entry under title:

Chemical taxonomy, molecular biology, and function of
plant lectins.

(Progress in clinical and biological research ; v. 138)
Includes bibliographies and index.
1. Plant lectins--Congresses. 2. Plant chemotaxonomy--
Congresses. 3. Molecular biology--Congresses. I. Goldstein, Irwin Joseph. II. Etzler, Marilynn E. III. E·Y
Laboratories. IV. Series.
QK898.L42C53 1983 582.13'0419245 83-19937
ISBN 0-8451-0138-2

Contents

Contributors

Anthony K. Allen, Department of Biochemistry, Charing Cross Hospital Medical School, University of London, Hammersmith, London, England **[71]**

Marilyn A. Anderson, Plant Cell Biology Research Centre, School of Botany, University of Melbourne, Parkville, Victoria, Australia **[143]**

Joseph W. Becker, Department of Developmental and Molecular Biology, The Rockefeller University, New York, NY **[31]**

Carl A.K. Borrebaeck, Department of Pure and Applied Biochemistry, University of Lund, Lund, Sweden **[251,253]**

Ralph A. Bradshaw, Department of Biological Chemistry, California College of Medicine, University of California, Irvine, CA **[255]**

Maarten J. Chrispeels, Department of Biology, University of California, San Diego, La Jolla, CA **[269]**

Adrienne E. Clarke, Plant Cell Biology Research Centre, School of Botany, University of Melbourne, Parkville, Victoria, Australia **[143]**

Bruce A. Cunningham, Department of Developmental and Molecular Biology, The Rockefeller University, New York, NY **[31]**

Frank B. Dazzo, Department of Microbiology and Public Health, Michigan State University, East Lansing, MI **[209]**

M. de Guia Silva Lima, Biochemistry Department, Universidade Federal Do Ceara, Fortaleza (CE), Brazil **[271]**

Marie-Thérèse Esquerré-Tugayé, Centre de Physiologie Végetalé-LA 241 CNRS, Université Paul Sabatier, Toulouse, France **[177]**

Marilynn E. Etzler, Department of Biochemistry and Biophysics, University of California, Davis, CA **[1,197]**

A. Foriers, Department of Medical and Special Biochemistry, Pharmaceutical Institute, Free University, Brussels (V.U.B.), Brussels, Belgium **[7]**

Günter Gebauer, Institut für Pharmazie und Lebensmittelschemie, Universität Würzburg, Würzburg, Bavaria, W. Germany **[237]**

Cristina Glad, Department of Pure and Applied Biochemistry, University of Lund, Lund, Sweden **[251,253]**

Irwin J. Goldstein, Department of Biological Chemistry, University of Michigan, Ann Arbor, MI **[21,131]**

F. Greer, The Rowett Research Institute, Bucksburn, Aberdeen, Scotland, UK **[271]**

Kenneth D. Hapner, Department of Chemistry, Montana State University, Bozeman, MT **[255]**

John J. Hemperly, Department of Developmental and Molecular Biology, The Rockefeller University, New York, NY **[31]**

The number in brackets is the opening page number of the contributor's article.

Eliot Herman, Department of Biochemistry, University of California, Riverside, CA **[259]**

Leslie M. Hoffman, Biochemistry Department, Agrigenetics Corporation, Madison, WI **[279]**

Rosslyn D. Hoggart, Plant Cell Biology Research Centre, School of Botany, University of Melbourne, Parkville, Victoria, Australia **[143]**

Roswitha Hopkins, Department of Botany and Plant Pathology, Oregon State University, Corvallis, OR **[117]**

Yasuyuki Imai, Division of Chemical Toxicology and Immunochemistry, Faculty of Pharmaceutical Sciences, University of Tokyo, Tokyo, Japan **[281]**

Heinrich Kauss, Biology Department of the University, Kaiserslautern, Federal Republic of Germany **[187]**

K. Keegstra, Department of Botany, University of Wisconsin, Madison, WI **[163]**

David C. Kilpatrick, Regional Blood Transfusion Centre, Royal Infirmary, Edinburgh, UK **[63]**

T.P. King, The Rowett Research Institute, Bucksburn, Aberdeen, Scotland, UK **[271]**

Rozenn N. Kouchalakos, Department of Biological Chemistry, University of California, California College of Medicine, Irvine, CA **[255]**

Jamie E. Lamb, Department of Pathology, University of Utah, Salt Lake City, UT **[21]**

M. Lauwereys, Department of Protein Chemistry, Institute of Molecular Biology, Free University of Brussels (V.U.B.), Sint Genesius-Rode, Belgium **[7]**

Thomas N. Metcalf, III, Department of Biochemistry, Michigan State University, East Lansing, MI **[273]**

M.L. Mishkind, Department of Botany, University of Georgia, Athens, GA **[163]**

Kenneth W. Olsen, Department of Chemistry, Loyola University of Chicago, Chicago, IL 60626 **[261]**

Toshiaki Osawa, Division of Chemical Toxicology and Immunochemistry, Faculty of Pharmaceutical Sciences, University of Tokyo, Tokyo, Japan **[281]**

B.A. Palevitz, Department of Botany, University of Georgia, Athens, GA **[163]**

Willy J. Peumans, Department of Plant Biochemistry, Katholieke Universiteit Leuven, Leuven (Heverlee) Belgium **[99]**

Sándor Pongor, Department of Medical Biochemistry, The Rockefeller University, New York, NY **[277]**

Indravathamma Poola, Department of Biological Chemistry, University of Michigan, Ann Arbor, MI **[21]**

A. Prouvost-Danon, Immunotherapie Experimentale, Institut Pasteur, Paris, France **[271]**

Steven G. Pueppke, Department of Plant Pathology, University of Florida, Gainesville, FL **[225]**

A. Pusztai, The Rowett Research Institute, Bucksburn, Aberdeen, Scotland, UK **[271]**

Ralph S. Quatrano, Department of Botany and Plant Pathology, Oregon State University, Corvallis, OR **[117]**

Natasha V. Raikhel, Department of Botany, University of Georgia, Athens, GA **[117,163]**

Zoltán Riedl, Department of Product Development, HUMAN, Institute for Serobacteriological Research, Budapest Hungary **[277]**

David D. Roberts, Department of Biological Chemistry, University of Michigan, Ann Arbor, MI **[21,131]**

Pierre Rougé, Département de Biologie cellulaire, Faculté des Sciences Pharmaceutiques, Université Paul Sabatier, Toulouse, Hte Garonne, France **[265]**

Harold Rüdiger, Institut für Pharmazie und Lebensmittelchemie, Universität Würzburg, Würzburg, Bavaria, W. Germany **[237]**

Melvin Schindler, Department of Biochemistry, Michigan State University, East Lansing, MI **[273]**

Johann Schurz, Institut für Pharmazie und Lebensmittelchemie, Universität Würzburg, Wurzburg, Bavaria, W. Germany **[237]**

Leland M. Shannon, Department of Biochemistry, University of California, Riverside, CA **[47]**

John E. Sherwood, Department of Microbiology and Public Health, Michigan State University, East Lansing, MI **[209]**

Hetty M. Stinissen, Department of Plant Biochemistry, Katholieke Universiteit Leuven, Leuven (Heverlee), Belgium **[99]**

A. Donny Strosberg, Department of Molecular Immunlogy, Institut J. Monod, CNRS and University Paris VII, Paris, France **[7]**

Alessandro Vitale, Department of Biology, University of California, San Diego, La Jolla, CA **[269]**

Lila O. Vodkin, Seed Research Lab, U.S. Department of Agriculture, Beltsville Agriculture Research Center, Beltsville, MD **[87]**

John L. Wang, Department of Biochemistry, Michigan State University, East Lansing, MI **[273]**

Takeshi Yamazaki, Division of Chemical Toxicology and Immunochemistry, Faculty of Pharmaceutical Sciences, University of Tokyo, Tokyo, Japan **[281]**

David H. Young, Department of Plant Pathology, University of Wisconsin-Madison, Madison, WI **[187]**

Chemical Taxonomy, Molecular Biology, and Function of Plant Lectins, pages 1–5

INTRODUCTION

Marilynn E. Etzler

Department of Biochemistry & Biophysics
University of California
Davis, CA 95616

The discovery of a lectin in castor beans by Stillmark in 1888 opened the door to a century of exploration of these fascinating molecules. We now know that lectins are present in a wide variety of plant species spanning almost every taxonomical classification in the plant kingdom. Indeed, as we develop more methods of detecting these carbohydrate-binding molecules, we are finding that many plants contain more than one lectin. Yet, despite the wealth of information that now exists on the specificities and structures of these molecules and the many applications of lectins to research in the areas of complex carbohydrates and cell surfaces, we do not yet have a thorough understanding of the relation of structure to function of these molecules and we know little about the regulation of their production or what biological role these lectins serve in the plant. At this meeting we have assembled a group of investigators who share a common interest in plant lectins yet have come to this area from a variety of scientific backgrounds and utilize a variety of approaches in our studies of these molecules. It is hoped that through discussions of our work and presentations of our various viewpoints we might develop new insights that may eventually lead to the elucidation of the role of these lectins in nature.

At present over 100 different plant lectins have been partially characterized with respect to their structures and specificities. Some of the most intensively studied lectins are listed in Table 1 and are categorized with respect to the taxonomical grouping of the plant from which

TABLE 1. TAXONOMICAL GROUPINGS OF SOME PLANTS CONTAINING WELL-CHARACTERIZED LECTINS

Leguminosae	Euphorbeaceae
Arachis hypogaea	*Abrus precatorius*
Canavalia ensiformia	*Ricinus communis*
Dolichos biflorus	Solanaceae
Glycine max	*Datura stramonium*
Griffonia simplicifolia	*Lycopersicon esculentum*
Lens culinaris	*Solanum tuberosum*
Phaseolus lunaus	Graminaceae
Phaseolus vulgaris	*Secale cereale*
Pisum sativum	*Triticum vulgaris*
Vicia faba	

they are derived. As can be seen in this Table, we have the most information about legume lectins which have proven an excellent source of material because of their abundance in the seeds of these plants. However as we continue to get more information on lectins from other families of plants it is becoming increasingly apparent that different classes of plants contain lectins with different molecular properties. For example, lectins from plants in the Solanaceae family are characterized by their high carbohydrate content and the presence of hydroxyproline residues; many of the legume lectins are rich in aspartic acid and serine and have little or no methionine or cysteine whereas the monocot lectins studied to date are basic proteins and have a large number of cysteine residues. It is important to bear in mind that although these different categories of lectins represent the most abundant species of lectins in their respective plants, they are not derived from homologous tissues. Therefore in the examples above, we are comparing lectins from the cotyledons of legumes with lectins from such organs as modified stems of the Solanaceae and from the embryos of the monocots. We must remember these differences in origin while making our comparisons and consider the possibility that different categories of plants as well as different tissues within the same plant may contain lectins that have been adapted to perform different functions.

Partial or complete amino acid sequences are now available for a number of lectins, making it possible to consider the relationships of these molecules to one another. Drs. Strosberg, Goldstein and Cunningham will be talking about these comparisons in the first session of this meeting. In this same session Drs. Shannon and Kilpatrick will discuss their studies in which they have used an immunochemical approach to determine the structural relationships among these molecules. Such studies are invaluable in determining the evolutionary and ultimately the functional relationships among the lectins.

In addition to comparing lectins with one another, it is of interest to compare different parts of the same lectin molecule, particularly with respect to their possible functional role. In the second session Dr. Allen will be talking about the domains of the potato lectin, and I am sure a number of speakers throughout this meeting will be addressing the relationship of structure to function of various lectins.

The advent of genetic engineering has recently made it possible to clone the lectin genes or obtain large amounts of DNAs complementary for the mRNAs of some lectins. We are now beginning to see a rapid proliferation of information on the nucleotide sequences of these DNAs from which we can readily obtain the complete amino acid sequences of the respective lectins. In the second session Dr. Vodkin will describe her work on the structure of the soybean lectin gene, and both she and Dr. Peumans will discuss some of the exciting work they are doing on the molecular biology of the soybean and cereal lectins. In the third session, Dr. Quatrano will address the question of the regulation of lectin synthesis and discuss his evidence in the case of wheat germ agglutinin that hormonal regulation may occur at the transcriptional level.

Although lectins are detected on the basis of their carbohydrate-binding properties, it is possible that these. molecules may also have other activities associated with them as has been found in the case of immunoglobulins. Of great interest in this respect has been the discovery of hydrophobic binding sites on a number of lectins that are separate and independent of the carbohydrate binding sites. David Roberts will present the studies which he and Dr. Goldstein are conducting on these hydrophobic binding sites and show that cytokinins can interact with these sites. The interaction of this class of plant hormones with lectins may provide some clue as to their function or to the regulation of activity of these molecules.

Essentially two types of approaches have been used in an effort to obtain information on the functional roles of lectins. The first approach has been to study the distribution and properties of lectins during the life cycle of the plant. Such basic information is essential before one can definitely ascribe any functional role to a lectin since the lectin would have to be present at the right place and the right time for its participation in that function. The second approach has been to try to establish whether lectins may be involved in any particular hypothetical roles. It is clear that a combination of both approaches is needed before we can answer the question of the function of lectins.

Several of us here at this conference have utilized the first approach, and Drs. Pueppke, Mishkind and I will be discussing our work in this area. One thing that has become apparent from this approach is that lectins may exist in various tissues of the same plant and have different cellular localizations and molecular properties. It is thus essential to distinguish these lectins from one another and recognize the possibility that these lectins may have different functions.

With respect to the second approach, a wide variety of roles have been proposed for plant lectins over the past 30 years. Some of these roles are summarized in Table 2. Several of these roles concentrate on seed lectins, such as the possibility of serving as storage proteins or in the packaging or mobilization of these materials. Another possible function for lectins that has been frequently mentioned in the literature is their involvement in various defense roles in the plant. Such

TABLE 2. SOME PROPOSED FUNCTIONS FOR LECTINS IN PLANTS

Storage proteins
Packaging and/or mobilization of storage materials
Transport of carbohydrates
Defense mechanisms
Attractants for Rhozobial symbiosis
Mitogenic stimulators of plant embryonic cells
Cell recognition
Cell wall extension
Enzymes

roles have included their possible involvement as fungistatic or bacteriostatic agents during germination and early growth, a subject that I am sure Dr. Mishkind will address in his presentation, and their possible function as receptors for elicitors which Dr. Albersheim may mention. Dr. Esquerre-Tugaye will consider the possible involvement of hydroxyproline rich glycoproteins and lectins in the defense of plants.

Considerable attention has been paid in recent years to the possible involvement of lectins in the symbiosis between Rhizobia and legumes. There has been quite a bit of controversy on this topic and Drs. Dazzo and Pueppke will discuss data that have led to different viewpoints on this subject. One particularly attractive functional role that has been proposed for lectins is their involvement in cell-cell interactions. Dr. Clarke will address this possibility in her talk in Session III.

The possibility that lectins may be involved in cell wall extension was raised by Dr. Kauss in 1974. Both Dr. Kauss and I will present our latest information that supports this possibility, and Dr. Esquerre-Tugaye will compare the properties of hydroxyproline rich cell wall glycoproteins and lectins.

Two of our speakers have in the past been strong proponents of the view that lectins serve as enzymes. I am sure that Dr. Albersheim will address this question in his talk and it will be of interest in the discussion to hear Dr. Shannon's latest views on this subject.

Finally, no discussion of lectin function would be complete without a consideration of what type of molecules serve as endogenous receptors for the lectins in the plants. This area was largely ignored until the very interesting work of Dr. Rudiger and we all look forward to hearing his recent findings on this subject.

We would be unrealistic to think that this Symposium is going to provide any concrete solutions to the problems addressed above. However through the presentations and and discussions of our various research efforts at this meeting we may develop new insights that direct our future research in this area. We thank E·Y Laboratories for having the foresight to arrange such a meeting and know that our research will profit from it.

Chemical Taxonomy, Molecular Biology, and Function of Plant Lectins, pages 7–20

MOLECULAR EVOLUTION OF LEGUME LECTINS

A. Donny Strosberg,

Molecular Immunology, Institut J. Monod,
C N R S and University Paris VII, 2 place Jussieu,
75251 Paris Cedex 05, France

M. Lauwereys,

Protein Chemistry, Inst. Molecular Biology,
Free University of Brussels (V.U.B.),
65 Paardenstraat, Sint Genesius-Rode, B 1640 Belgium

A. Foriers,

Medical and Special Biochemistry, Pharmaceutical Institute, Free University Brussels (V.U.B.),
103 Laarbeeklaan, B 1090 Brussels, Belgium

INTRODUCTION

Lectins constitute a diverse group of proteins that share the ability to specifically bind carbohydrates, in particular on glycoproteins and glycolipids (Lis and Sharon, 1977). While these proteins have been found in a large variety of organisms ranging from bacteria to vertebrates, the study of their structure has most progressed for the plant lectins, where they were first discovered. Purified from the seeds of the leguminosae, they were initially used in blood typing, since different lectins agglutinate erythrocytes bearing different carbohydrates. The discovery that their binding to glycoproteins of lymphocytes might trigger cell mitosis led to the extensive use of some lectins in immunology and hematology. However, the wide utilization of lectins has done little for the elucidation of their physiological function in plants. For this purpose, we and others have initiated the comparative study of the legume lectins of different origins with the aim of unravelling the principles that govern structure-function relationship in this fascinating group of proteins.

COMMON GENERAL PROPERTIES

Initial studies suggested that lectins extracted from seeds of leguminous plants differed by a variety of molecular properties including size, charge, heterogeneity and, most proeminently, the structure of the sugar moeity which they were able to recognize (Lis and Sharon, 1977).

As more information emerged it became apparent that the legume lectins were closely related. Indeed, most are composed of subunits with a molecular weight of 25,000 to 30,000, associated as dimers of about 50,000 or tetramers of about 100 to 120,000 daltons. The subunits may be constituted by a single polypeptide chain or by two chains of different size.

Recent studies of the *in vitro* translated products of mRNA coding for pea, lentil and fava bean lectins confirmed the hypothesis (Foriers et al., 1978) that these two chain-proteins are initially synthetized as a single polypeptide which is post-synthetically cleaved into the larger β and the smaller α chains (Hemperley et al., 1982; Lauwereys et al., 1983).

We present in figure 1 the distribution of the leguminous plants from which lectins have been sequenced and are compared in this report.

AMINO ACID SEQUENCE STUDIES

Amino terminal region

The sequence determination of the amino terminal portion of the lectin polypeptide chains initially yielded confusing results, since Con A and the α and β chains of the lentil and pea lectins appeared as very different. However, since the complete sequence of Con A was already available, it soon became apparent that these legume lectins, which all three bind mannose and glucose, also shared homologous sequences albeit in different portions of the molecule (Figure 2).

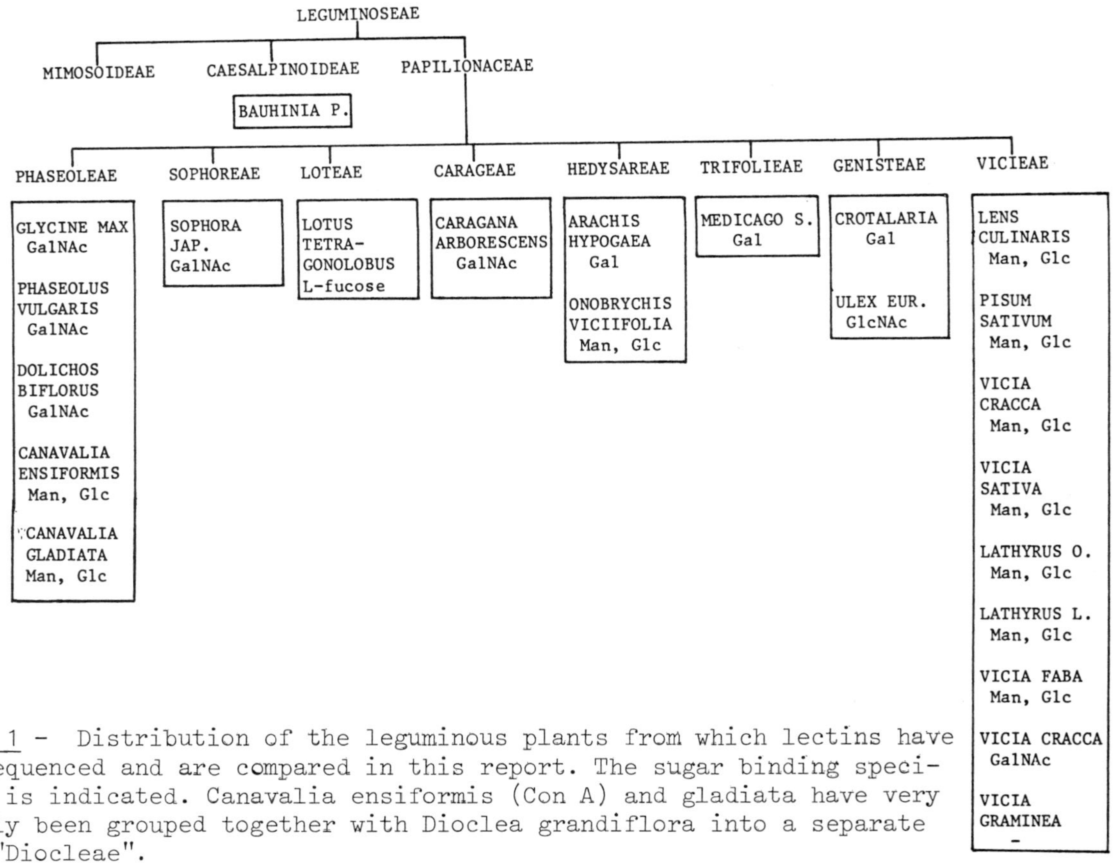

Figure 1 - Distribution of the leguminous plants from which lectins have been sequenced and are compared in this report. The sugar binding specificity is indicated. Canavalia ensiformis (Con A) and gladiata have very recently been grouped together with Dioclea grandiflora into a separate tribe "Diocleae".

```
                              123                                130                                          140
Con A                         Thr-Asp-Ala-Leu-His-Phe-Met-Phe-Asn-Gln-Phe-Ser-Lys-Asp-Gln-Lys-Asp-Leu-Ile-Leu-Gln-Gly-Asp-Ala-Thr-

                              1                                        10                                   20
Crot.J.                       Ala-Glu-Glu-Gln-Ser-Phe-Ser-Ser-Thr-Lys-Phe-Ser-Thr-Asp-Gln-Pro-Asn-Leu-Ile-Leu-Gln-Gly-Asp-Ala-Thr-
Sophora J.                    ——— Ile-Leu ———— Phe-Pro ——— Ala-Ser-Asn-Glu-Thr-Asp —— Leu ——————————— Leu-
Caragana A.                   Thr —— Ser-Leu ———— Phe-Asn ——— Val-Pro-Asn ——— Asp ———— His —— Glu —— Leu-
Ulex E.                       Ser-Asp-Asp-Leu ——— Lys —— Lys-Asn ——— Gln-Asn-Gly-Lys-Asp —— Ser-Phe ——— Ser —— ( )
Bauhinia P.              Thr-Ser-Ser-Thr-Leu-Thr ——— Thr-Phe-Pro-Asn —— Trp-Ser-Asn-Thr-Gln-Glu —— Gly-Thr-Ser-Ile-Ile-Phe-Gln-
V.Gram                                                                     Ser-Gly-Ile —— Phe ———————— Ser-
Medicago S.                   ——— Ser-Glu ———— Phe —— Ser —— (                                                )
PNA                           ——— Thr-Val ——— Asn-Phe-Asn-Ser ——— Glu-Gly-Asn —— Ala-Ile-Asn-Phe ——— Val ——
SBA                           ——— Thr-Val ———— Trp-Asn ——— Val-Pro-Lys-Glu —— Asn-Met ——— Glu ———— Ile-
V.Cracca                      Thr —— Ser-Thr ———— Phe —— Glu —— Asn-Gln-Asn —— Gln ———— Glu ————————
Dolichos B.                   —— Asn-Ile ————— Phe-Lys-Asn —— Asn-Ser [     ] —— Ser-Phe ———————————
Lotus T.                              Val ———— Tyr —— Glu —— [ ] Lys —— Asp-Gly-Ser ———————— Arg —— Lys-
Phas.R                        —— Ser —— Thr ———— Phe-Glu-Arg —— Asn-Glu-Thr [     ] ———————— Arg ——— Ser-
Pinto III                     —— Ser-Gln-Thr-Phe ——— Phe-Asp-Arg —— Asn-Glu-Thr [     ] ———————— Gly ——— Ser-
Phas.L                        Ser-Asn-Asp-Ile-Tyr —— Asn-Phe-Glu-Arg —— Asn-Glu-Thr [     ] ———————— Arg ——— Ser-
Lentil β                      Thr —— Thr-Thr ———— Ile ————— Pro ——— Gln ———— Phe ———— Gly-Tyr-
Pea β                         Thr —— Thr-Thr ——— Leu-Ile ————— Pro ——— Gln ———— Phe ——— Asn-Gly-Tyr-
V.Cracca β                    Thr —— Thr-Thr ——— Leu-Ile ————— Pro ——— Gln ———— Phe ———— Gly-Tyr-
Favin β                   Thr-Asp —— Ile-Thr ———— Ile-Pro ——— Arg-Pro ——————————— Phe ——— Gly-Gly-Tyr-
V.Sativa β                    Thr —— Thr-Thr ——— Leu-Ile ————— Pro ——— Glu ———— Tyr ———— Gly-Tyr-
Lath.O. β                     Thr —— Thr-Thr ——— Leu-Ile ————— Pro ——— Gln ———— Phe ———— Gly-Tyr-
Lath.S. β                     Thr —— Thr-Thr ——— Leu-Leu ———— Gly-Pro ——— Gln ———— Phe ———— Gly-Tyr-
Sanfoin I                 Ala-Glu-Asn-Thr-Val ——— Asp-Phe-Ser ——— Leu-Ser-Gly —— Glu-Asp ———————— Thr-Val-
Sanfoin II                Ala-Glu-Asn-Ile-Val ——— Asp-Phe-Ser ——— Leu-Ser-Gly —— Glu-Asp ———————— Thr-Val-
```

Figure 2 - N-terminal amino acid sequences of 25 lectins from leguminous plants, compared to a homologous portion 123 to 147 from Con A. The protein from Crotalaria juncea was chosen as the model and homologies are indicated by a line. Brackets indicate a deletion introduced to maximize the homologies, and parentheses indicate unknown residues. References for the sequence data are : for Con A (the lectin of Canavalia ensiformis) : Edelman et al., 1972; Hardman and Ainsworth, 1972; for Crot. J. (Crotalaria juncea) and Lotus T. (tetragonolobus) : Foriers et al., 1979; for Sophora J., Caragana A., Ulex E., Bauhinia P., V. Graminea and Medicago Sat. : Lauwereys et al., 1983; for PNA (Peanut Agglutinin, the lectin from Arachis Hypogaea) and SBA (Soy Bean Agglutinin, the lectin from Glycine Max) : Foriers et al., 1977b; for V. Cracca : Baumann et al., 1979; for Dolichos B. : Etzler et al., 1977; for Phas (Phaseolus) L. and Phas R. : Miller et al., 1973; for Pinto III (another kind of Phaseolus vulgaris) variety of PHA : Lauwereys et al., 1983; for the β chains of the two-chain lectins the references are : for Lentil (Lens culinaris) : Van Driessche et al., 1977a; for pea (Pisum sativum) : Van Driessche et al., 1976; for V. Cracca : Baumann et al., 1979; for Favin : Cunningham et al., 1979; for V. Sativa : Gebauer et al., 1979; for Lath. O. (Lathyrus Odoratus) and Lath. S. (Sativa) : Kolberg et al., 1980, 1982; for Sanfoin I and II (the lectins from Onobrychis viciifolia) : Young et al., 1982.

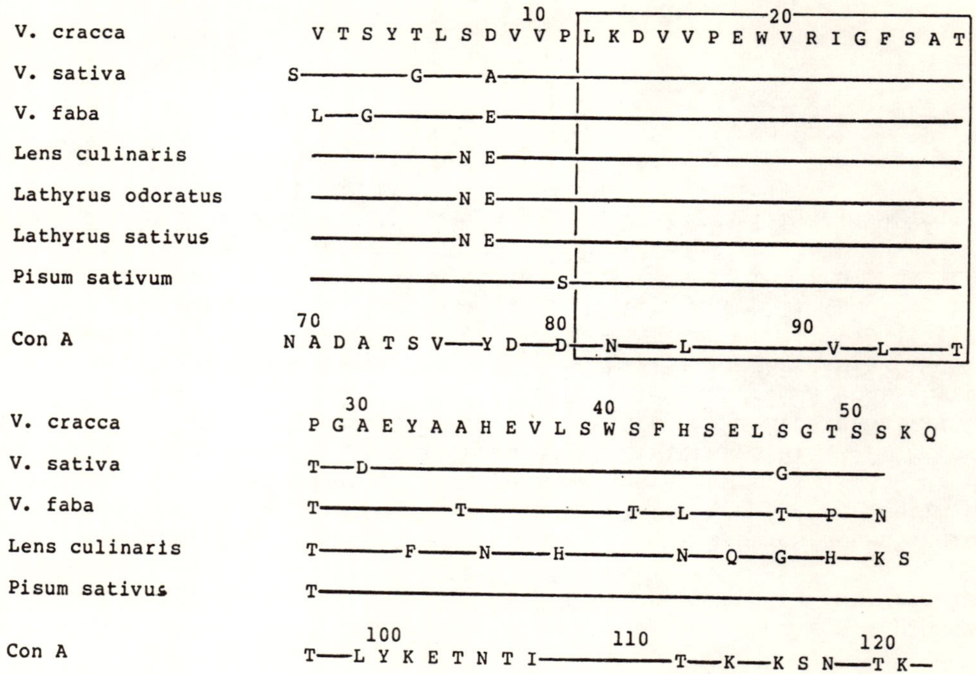

Figure 3 - Amino acid sequences of the α chains from two-chain lectins. The V. cracca α chain was chosen as a model, and sequence identities are indicated by a line. We have enclosed in a box a completely identical portion of the chain between positions 12 and 27. Refs. are as in Fig. 2.

Analysis of the N-terminal portions of many other lectins purified since, have indeed confirmed that Con A was the only exception : all other lectins were homologous in similar parts of their structures.

We grouped in Figures 2 and 3 the sequences obtained by various laboratories, according to the plant tribe to which the legumes belong and introduced deletions to maximize homology. In Figure 2 we grouped the N-terminal sequences of the one-chain lectins together with those of the β chains of the two-chain lectins. While some groups of lectins are closer to each other, all these proteins are similar in their N-terminal portions and a number of residues emerge in fact as invariant.

In Figure 3 we grouped the complete sequences of the α chains of the two-chain lectins compared to the homologous portions of Con A between positions 69 and 122.

Amino terminal sequences are available for lectins from non-leguminous plants, such as ricin, but no homology could be detected with the lectins compared here.

While the homology of amino terminal portions may be suggestive of further homology throughout the chains, only the determination of the complete sequences allow for a meaningful comparison. The whole primary structure of several lectins has now been determined, and their comparison (Fig. 3) fully supports and extends the conclusions drawn from the amino terminal sequence studies.

Complete sequences

The complete sequence of favin (Cunningham et al., 1979) and lentil (Foriers et al., 1981) lectin have been determined by classical methods using the protein sequenator and various proteolytic cleavages. These sequences were compared to that of Con A previously determined by similar methods. To this comparison one now can add the recently published sequence of the single polypeptide lectin from French bean (Phaseolus vulgaris) deduced from the nucleotide sequence of a cDNA plasmid prepared by reverse transcription of mRNA (Hoffman et al., 1982).

As predicted from the studies on the two-chain lectins and their single polypeptide chain precursors, it appears that all leguminous lectins are homologous throughout their sequence, provided one introduces appropriate deletions.

The amino acid sequences of the β and α chains of favin and lentil lectins, of Con A and PHA are shown in Figure 4. The similarities are extensive; but to maximize the homology, the Con A sequence had to be rearranged into two portions : one (positions 122 to 237 and 1 to 69) aligned with positions 1 to 183 of the favin β, 1 to 159 of the lentil β lectin and 1 to 154 of PHA, the other (positions 70 to 121) aligned with positions 1 to 51 or 52 of the favin and lentil lectin α chains and positions 155 to 209 of PHA.

Metal binding sites

In Con A the metal binding site is situated between positions 1 to 45. It appears that this region is highly conserved between the various lectins compared here. Thus of the 45 residues, 24 are identical between the favin lentil, lectin and Con A, and all the amino acids which interact directly with the metal ions in Con A are homologous in the favin and lentil lectin. In PHA 18 residues are identical. In fact, favin and LcL only differ at six out of the 45 positions, thus stressing the highly conserved character of this portion of the proteins.

In Con A, the Ca^{2+} binding site contains residues Asp^{10}, Tyr^{12} Asp^{14} and Asp^{19} : the same residues appear in homologous positions in other lectins except for the replacement of Tyr^{12} by another aromatic residue, phenylalanine at the corresponding position 125 in favin, LcL and PHA. In this last protein, however, Asn^{14} is replaced by a Ser residue and Asp^{19} is deleted. The Mn^{2+} binding site in Con A is constituted by the side chains of Glu^{8}, Asp^{10}, Asp^{19} and His^{24}. In favin and in LcL all these residues are present in equivalent positions. In PHA the first two are identical, Asp^{19} is missing and His^{24} is replaced by another basic residue, arginine.

It is worth noting that most of the residues involved in metal binding in Con A also appear to be present in SBA (Vodkin et al., 1983) and in PNA (Lauwereys et al., in preparation).

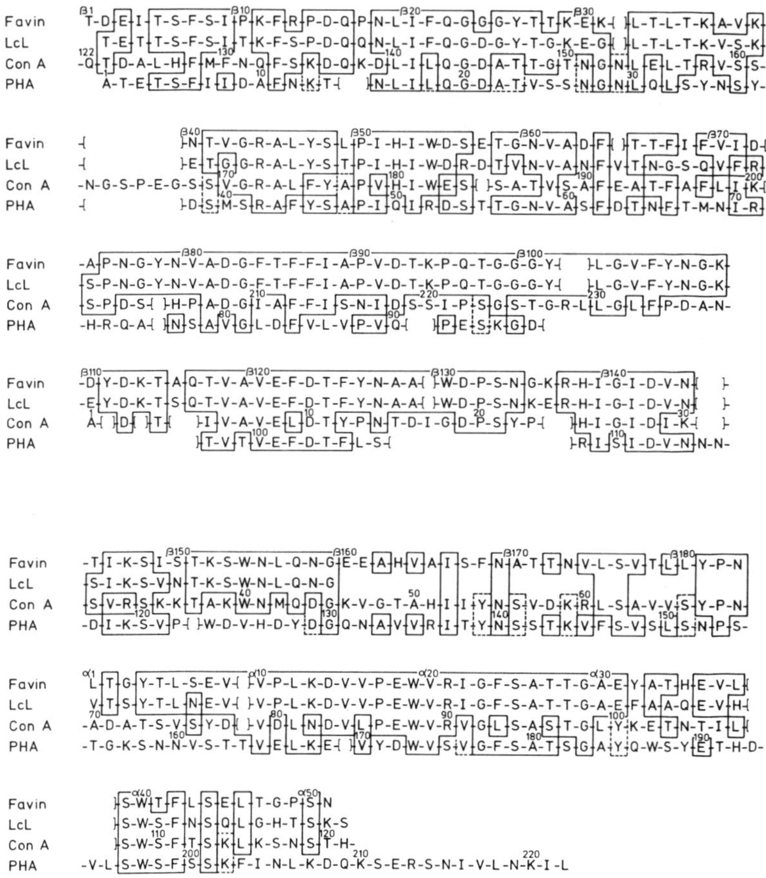

Figure 4 - Complete sequences of favin, LcL, Con A and PHA. The identical residues have been boxed in. Empty brackets correspond to deletions. The empty space at the end of the LcL β, and the favin, and LcL α chains may denote post-synthetic removal of small fragments. The references are given in the text.

Hydrophobic cavity

Of striking significance is the conservation in LcL, favin, and PHA of the amino acid residues which contribute to the three-dimensional structure of the well characterized hydrophobic cavity of Con A. Except for Tyr^{54}, which is present in PHA but replaced by the homologous Phe in favin and absent in LcL, all the other hydrophobic residues which form the Con A cavity described by Becker et al. (1975) are perfectly aligned in favin and in LcL. In addition, the Con A residue Leu^{85} is a Val in all three other proteins, Val^{91} is Val in PHA, but Ile in LcL and favin, and finally Val^{179} is Ile in all three other lectins.

The conservation of this hydrophobic cavity through evolution supports its essential role in the function of plant lectins, especially since it has been shown that Con A can bind plant auxins (Edelman and Wang, 1978).

Glycosylation site

The favin β chain contains a carbohydrate moiety attached to a characteristic Asn-X-Thr/Ser sequence. The -Asn^{169}-Ala-Thr^{171} sequence is absent in LcL and is replaced by -Asn-Ser-Val in Con A also devoid of covalently bound sugar, but by -Asn^{140}-Ser^{141}-Ser^{142} in PHA, in which it has been predicted to be one of four glycosylation sites, the others being at positions 12 to 14, 65 to 67, 152 to 154 and 160 to 162 (Miller et al., 1975; Hoffman et al., 1982).

Carbohydrate binding site

The amino acid residues which constitute the sugar binding site in Con A (Hardman and Ainsworth, 1976; Becker et al., 1976) are poorly conserved in LcL and favin : only one-third of the relevant positions are occupied by homologous residues in these lectins. This divergence is surprising considering that Con A, LcL and favin all three bind mannose and glucose, albeit with variations in association constants (Hörejsi et al., 1977). More work is needed to characterize the sugar binding sites of the lectins compared here.

Folding and three dimensional structure

The striking homologies between the sequences of the one and two-chain lectins is strongly suggestive of a good ressemblance of the folding and three-dimensional structure of these proteins. The secondary structure of Con A based on X-ray crystallographic determination, and the probable secondary structure of the LcL predicted by two methods of calculation appeared essentially similar (Foriers et al., 1981) : the large majority of β pleated sheets found in Con A are present in homologous positions in LcL, with the same connecting order, except where deletions occurred. The β-turn regions are localized similarly. Metals contribute to the structural stability of the lectins : the octohedral metal binding sites discussed already in terms of similar amino acid sequences, are most likely to retain the same configuration.

Whether constituted of one or of two polypeptide chain, and regardless of the fact that the amino terminal 72 residues of Con A are homologous to the carboxy-terminal residues of the β-chain, it appears that the legume lectins also share common three-dimensional structures (Olsen, 1982).

Molecular Genetics

In comparing the various legume lectin sequences now available, the striking fact remains that a region that is carboxy-terminal in most lectins became amino terminal in Con A. The most likely explanation for this rearrangement of genetic material involves the existence at one time in evolution of a duplicated pair of genes coding for a single polypeptide chain of an approximate length of 240 residues (Fig. 5). Three protein coding nucleotide sequences are distinguished : β_1 corresponds to the first 109 residues of the β chains of the two-chain lectins and of the one-chain lectins others than Con A and is equivalent to positions 122 to 237 in Con A, β_2 corresponds to residues 110 to 159 in the β-chain of LcL, to 110 to 183 in the β-chain of favin and in the one-chain lectins, and is the region 1 to 69 in Con A. Finally, α is the small chain of favin, LcL and other two-chain lectins, is the C-terminal portion of the one-chain lectins and is region 70 to 122 in Con A. Non coding nucleotide sequences ("introns"), depicted as loops in Fig. 5, may separate the coding stretches of DNA, to facilitate alternative rearrangements possibly used during cell differentiation and expression of lectin genes in various organs of the plant. After transcription and translation the β_1 - β_2 - α precursor polypeptides preceeded by an hydrophobic leader sequence (not shown in the figure) is processed

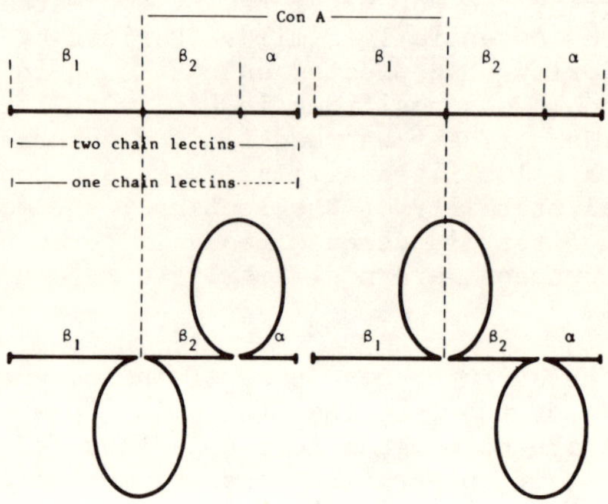

Figure 5 - Model for the gene organization of lectins. According to this model, the ancestor gene coding for all leguminous plant lectins is arranged in the $\beta_1 - \beta_2 - \alpha$ order, and Con A results from rearrangements involving duplicated genes.

into α and β chains with possible loss of additional short segments which would explain the size differences between the favin and LcL β and β' chains (Foriers et al., 1981; Cunningham et al., 1979). The Con A gene would have resulted from the use of genetic material of two adjacent genes.

The hypothesis of gene duplication is strongly supported by the discovery of both a one- and a two-chain lectins with different amino terminal sequences in a single plant, V. cracca (Baumann et al., 1979). The duplication of the SBA gene has been reported. Moreover, the existence of several isolectins in various legumes also argues in favor of the presence in the genome of several genes.

Proteolytic maturation of lectins

The comparison between the encoded information in the genes and the sugar binding lectin proteins clearly indicates that proteolytic cleavages modify a single polypeptide chain : the removal of a hydrophobic signal peptide is followed, for the two-chain lectins such as favin or pea lectin for instance, by a cleavage which yields the larger, amino terminal β-chain, and the smaller α-chain. Further "trimming" of these chains is evident by the difference in length between the β and β' chains of favin, and the β-chain of the lentil lectin, and by the apparent loss of 6 to 10 residues from the end of the α-chain as compared to the encoded information.

The one-chain lectins are also "trimmed". Thus the vicia graminea lectin is shorter by about 16 residues, and the lotus tetragonolobus lectin by about 4 residues when compared to the other lectins.

What is the role of this processing of the lectins ? Do the cleavages generate the carbohydrate binding properties or do they destroy an initial enzymatic activity by the precursor chain ? Do the same maturation steps occur for leaves and root than for seed lectins ? These questions may find an answer when sufficient unprocessed material will be made available, possibly by expression of cloned genes in cells which do not contain the processing enzymes. Introduction and expression of lectin genes in plants which do not make lectins should allow to investigate their role in the development of the plant, the regulation of its growth and the interaction with microorganisms.

ACKNOWLEDGEMENTS

The work reviewed here has been supported by grants from C.N.R.S. and a contract "Fixation de l'azote" funded by Elf-Aquitaine, E.M.C., C.D.F. Chimie and Rhône-Poulenc.

REFERENCES

Becker JW, Reeke GN, Cunningham BA, Edelman GM (1976). Nature 259:406.

Cunningham BA, Hemperly JJ, Hopp TP, Edelman GM (1979). Proc. Natl. Acad. Sci. USA 76:3218.

Edelman GM, Cunningham BA, Reeke GN Jr, Becker JW, Waxdal MJ, Wang JL (1972). Proc. Natl. Acad. Sci. USA 69:2580.
Etzler ME, Talbot CF, Ziaya PR (1977). FEBS Lett. 82:39.
Foriers A, De Neve R, Kanarek L, Strosberg AD (1978). Proc. Natl. Acad. Sci. USA 75:1136.
Foriers A, De Neve R, Strosberg AD (1979). Phys. Veget. 17: 597.
Foriers A, Lebrun E, Van Rapenbusch R, De Neve R, Strosberg AD (1981). J. Biol. Chem. 256:5550.
Foriers A, Van Driessche E, Wuilmart C, De Neve R, Kanarek L, Strosberg AD (1977a). FEBS Lett. 75:237.
Foriers A, Wuilmart C, Sharon N, Strosberg AD (1977b). Biochem. Biophys. Res. Commun. 75:980.
Gebauer G, Schiltz E, Schimpl A, Rüdiger H (1979). Hoppe-Seyler's Z. Physiol. Chem. 360:1727.
Hardman KD, Ainsworth CF (1972). Biochemistry 11:4910.
Hardman KD, Ainsworth CF (1976). Biochemistry 15:1120.
Hemperly JJ, Hopp TP, Becker JW, Cunningham BA (1979). J. Biol. Chem. 254:6803.
Hemperly JJ, Mostov KE, Cunningham BA (1982). J. Biol. Chem. 257:7903.
Hoffman LM, Ma Y, Barker RF (1982). Nucl. Acids Res. 10:7819.
Hörejsi V, Ticha M, Kocourek J (1977). Biochim. Biophys. Acta 499:290.
Lauwereys M, Shannon LM, Chemla R, Buffard D, Peumans W, Strosberg AD (1983). Submitted for publication.
Lauwereys M, Foriers A, Sharon N, Strosberg AD. In preparation.
Lis H, Sharon N (1977). In Sela M (ed): "The Antigens", Vol. 4, New York : Academic Press, p. 429.
Miller JB, Hsu R, Heinrikson R, Yachnin S (1975). Proc. Natl. Acad. Sci. USA 72:1388.
Olsen KW (1983). Biochim. Biophys. Acta 743:212.
Van Driessche E, Foriers A, Strosberg AD, Kanarek L (1976). FEBS Lett. 71:220.
Vodkin LO, Rhodes PR, Goldberg R (1983). Cell : in press.
Young NM, Williams RE, Roy C, Yaguchi M (1982). Can. J. Biochem. 743:212.

Chemical Taxonomy, Molecular Biology, and Function
of Plant Lectins, pages 21–29

STRUCTURAL HOMOLOGIES OF SOME LEAF AND SEED LECTINS

Irwin J. Goldstein, Jamie E. Lamb,
David D. Roberts, and Indravathamma Poola

Department of Biological Chemistry
The University of Michigan
Ann Arbor, Michigan 48109

Commencing with concanavalin A, the complete primary structure of 3 lectins (concanavalin A, the broad bean,(*Vicia faba*), and the lentil (*lens culinaris*) has been reported during the past decade. The partial amino acid sequence of a large number of other lectins is also available. These data have permitted phylogenetic relationships to be examined and a new field, the chemical toxonomy of plant lectins, to be established (Foriers *et al*.1979).

Investigations of the sequence of the legume lectins have proved to be especially rewarding. Thus the pea (*Pisum sativum*), the lectil (*Lens culinaris*), the broad lean (*Vicia faba*) and the *Vicia cracca* (mannose-binding) lectins, all of which contain two α- and two β-chains, exhibit pronounced homology among their α- and β-chains. The remarkable finding that the amino acid sequence of concanavalin A is circularly permuted relative to that of other legume lectins was reported by Foriers *et al*. (1978) for the lentil lectin and Cunningham and his colleagues (1979) and Hopp *et al*. (1982) for the *Vicia faba* lectin. Homology of legume lectins with concanavalin A is apparent when the amino termini of the β-chains of these lectins is aligned with residues 120-123 of concanavalin A and the α-chain with the region between positions 70-72 and 121 of the jack bean lectin (see Fig. 2 pg. 34).

For many years, our laboratory has been studying the lectins present in *Griffonia simplicifolia* seeds and leaves (Goldstein 1981). In this communication we report on some structural studies of the *G. simplicifolia* I (GS

I) and *G. simplicifolia* II (GS II) seed and leaf lectins. The isolation and characterization of a peptide from lima bean lectin and its homology with other mannose-binding lectins will also be discussed.

The GS I lectin consists of a family of 5 isolectins each of which is a tetrameric glycoprotein composed of varying proportions of two different subunits, A and B (Murphy, Goldstein 1977; Murphy, Goldstein 1979). The isolectins are designated A_4, A_3B, A_2B_2, AB_3 and B_4. The carbohydrate binding specificity of the subunits differs in that although both subunits bind α-D-galactopyranosyl end-groups the A subunit also binds 2-acetamido-2-deoxy-α-D-galactopyranosyl groups with high affinity (Goldstein *et al*. 1981; Murphy, Goldstein 1977; Murphy, Goldstein 1979).

Extensive homology between the A and B subunits was evident from the essentially complete cross reactivity of each subunit with antiserum raised against the other subunit. Additionally, the A and B subunits have closely similar amino acid analyses, differing in only one fundamental respect: The B subunit contains one methionyl residue whereas the A subunit contains none (Murphy, Goldstein 1977; Murphy, Goldstein 1979).

NH_2-terminal sequence analyses have been conducted on the A and B subunits. The discovery that both subunits were blocked at their NH_2-termini by pyroglutamate groups necessitated treatment with calf liver pyroglutamate aminopeptidase prior to primary structure analysis (Lamb, Goldstein, in press) Subsequent amino terminal sequence analysis of the GS I-A and -B subunits gave the results shown in Fig. 1. Note that there are only 4 differences between the 2 subunits in the sequence of the first 25 amino acid residues. These are at positions 8, 9, 17 and 19 which are Asn, Leu, Lys and Asn in subunit A, and Thr, Phe, Glu and Ser in subunit B. Homology of these GS I subunits with some galactosyl- and *N*-acetylgalactosaminyl-binding legume lectins is also tabulated in Fig. 1. The polypeptide chains are aligned to afford maximum homology. Note the phenylalanyl residues at positions 6 and 11; these residues appear to be highly conserved (cf Foriers et al 1979). It should be pointed out that although *Griffonia simplicifolia* is from the same family (Leguminoseae) as these plants it belongs to a different subfamily (Caesalpinoideae).

Treatment of the A and B GS I subunits with yeast carboxypeptidase Y released the same sequence of amino

		1				5					10					15					20					25
GS I-A	PyroGln	S	D	S	V	S	F	N	L	P	N	F	W	S	D	V	K	D	N	I	I	F	Q	G	D	A
GS I-B	PyroGln	S	D	S	V	S	F	T	F	P	N	F	W	S	D	V	E	D	S	I	I	F	Q	G	D	A
A. hypogaea		A	E	T	V	S	F	N	F	N	S	F	S	E	G	N	P	A		I	N	F	Q	G	D	V
D. biflorus		A	N	T	Q	S	F	S	F	Q	N	F	N	S	P	S	F			I	L		Q	G	D	A
G. max		A	E	T	V	S	F	S	W	N	K	F	V	P	K	E	P	D	M	I	L	E		G	D	A
PHA E		A	S	E	T	S	F	S	F	E	R	F	N	E	T			N	L	I	L		Q	R	D	A
L		S	N	D	I	Y	F	N	F	E	R	F	N	E	T			N	L	I	L		Q	R	D	A
V. cracca GalNAc		T	E	S	T	S	F	S	F	T	E	F	N	Q	N	Q	E	N	L	I	L	E		G	D	A
Con A	123	T	D	A	L	H	F	M	F	N	Q	F	S	K	D	Q	K	D	L	I	L		Q	G	D	A

Fig. 1. NH_2-Terminal sequences of some D-galactose- and N-acetyl-D-galactosamine-binding lectins.

acids, namely Arg-(Phe, Val)-Leu - Thr-Ser-COOH. These data indicate that rather similar sequences occur at both the amino and carboxyl termini of the two subunits. Digestion of the subunits with trypsin followed by peptide analysis by reverse-phase HPLC showed that approximately 30% of the tryptic peptides were common to both subunits. It was also shown that the cysteine-containing and glycosylated tryptic peptides of the A and B subunits had different retention times as analyzed by HPLC.

Using the procedure of Talbot and Etzler (1978) we have also isolated three lectins from G. simplicifolia leaves (Lamb et al 1983). The same affinity chromatographic columns employed to purify the seed lectins were used to isolate leaf lectins GS I, II and IV. No GS III lectin was found in G. simplicifolia leaves. Nor did rabbit antiserum raised against the seed GS III lectin cross react with leaf extracts (Poola and Goldstein, unpublished data).

Polyacrylamide gel electrophoretic analysis of the GS leaf lectins indicated that GS I and GS IV possessed the same subunit structure as the corresponding seed lectins. The hemagglutinating and carbohydrate binding activity of these leaf lectins also was similar to the corresponding seed lectins, and on the basis of these criteria, the lectins from these 2 sources are probably similar. However, the leaf GS II lectin gave a gel pattern quite distinct from the seed lectin. Thus the leaf lectin is composed of two types of subunits (Mr = 33,000 and 19,000) whereas the seed lectin consists of a single type of subunit (Mr = 32,500) (Lamb, Goldstein 1983). The two leaf lectin subunits have been separated by gel electrophoresis and the "slow subunit" (Mr = 33,000) was subjected to manual Edman sequencing. The preliminary NH_2-terminal sequence of the leaf "slow subunit" along with those of the GS II seed lectin and other legume sequences is presented in Fig. 2. Note that the seed GS II subunit and the leaf lectin "slow subunit" possess similar if not identical NH_2-terminal sequences. Precisely the same observation was made by Talbot and Etzler (1978) in their analysis of the CRM from Dolichos biflorus leaf and stems.

The "fast" subunit (M_r = 19,000) appears to be blocked at its NH_2-terminus. It would indeed be interesting if the blocking group were found to be pyroglutamate, the same group that was found at the NH_2-terminus of the GS I-A and -B subunits.

		1				5					10					15					20					25
GS I-A	PyroGln	S	D	S	V	S	F	N	L	P	N	F	W	S	D	V	K	D	N	I	I	F	Q	G	D	A
GS I-B	PyroGln	S	D	S	V	S	F	T	F	P	N	F	W	S	D	V	E	D	S	I	I	F	Q	G	D	A
GS II (Seed)		A	D	T	I	C	F	T	Y	T	N	F	G	L	D	V	S	D	L	S	L		Q	G	A	A
GS II (Leaf)		A	D	T	I/V	C	F	T	Y	T	N	F	G													
Lotus					^{1}V	S	F	S	Y	T	E	F		K	D	D	G	S	L	I	L		Q	G	R	A

Fig. 2. Comparison of the NH_2-terminal sequences of GS II seed and leaf lectins with other legume lectins.

Lima bean lectin was among the first plant agglutinin reported to be blood group specific (Boyd, Reguera 1949; Boyd, Shapleigh, 1954); it agglutinates human type A erythrocytes and binds N-acetyl-D-galactosamine. The lectin was isolated in pure form by Gould and Scheinberg (1970a) and Galbraith and Goldstein (1970; 1972). Lima bean lectin is composed of 31,000 dalton subunits each of which contains 2 cysteinyl residues. One cysteine is involved in an interchain disulfide linkage between 2 subunits to afford dimers (M_r = 62,000 daltons). Two non covalently-linked dimers comprise lima bean lectin component III (LBL III), and 4 such dimers give LBL II. Each subunit also contains a cysteinyl residue which has been shown to be essential for carbohydrate binding activity (Gould, Scheinberg, 1970b; Galbraith, Goldstein 1972). Fig. 3 depicts the anatomy of the lima bean lectins as described.

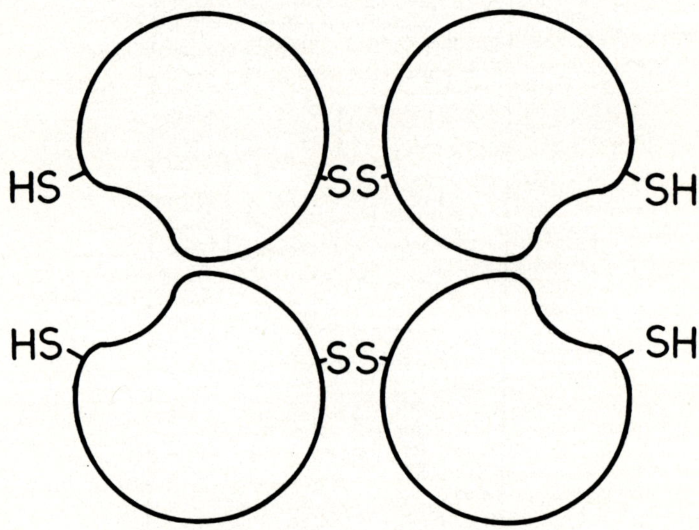

Fig. 3. Diagram of the subunit structure of lima bean lectin component III. Each subunit has M_r=31,000.

The position of the essential cysteinyl residue was determined to be approximately 1/3 of the amino acid residues distant from the carboxyl terminus of the LBL polypeptide chain. Peptides containing the free sulfhydryl group were generated by digestion with pepsin and trypsin followed by isolation on a mercurial agarose column (Roberts, Goldstein in press). Preliminary sequence analysis of the 3 major peptides gave a single common sequence:

Val-Glu-Phe-Asp-Thr-<u>Cys</u>-His-Asn-Leu-Asp-

Maximum homology with concanavalin A was obtained by aligning this LBL sequence with a sequence of the jack bean lectin commencing at valine 7. Sequence homology with concanavalin A, the pea, lentil and *Vicia faba* lectins is shown in Fig. 4. Major interest in this LBL sequence derives from the fact that the homologous amino acid sequence in concanavalin A contains residues which bind to Ca^{2+} and Mn^{2+} (Glu 8, Asp 10, Asn 14, Tyr 12) and make contact with carbohydrate in the concanavalin A sugar binding site (Tyr 12). Lima bean lectin is also a metaloprotein requiring Mn^{2+} and Ca^{2+} for its activity (Galbraith, Goldstein 1970). Currently, we are raising antibodies to these peptides in order to determine if we can assay for homologous sequences in other mannose-binding legume lectins.

LBL		V	E	F	D	T	C	H	N	L	D
Con A	7	V	E	L	D	T	Y	P	N	T	D
Favin β	119	V	E	F	D	T	F	Y	N	A	A
Lentil β	120	V	E	F	D	T	F	Y	N	A	A

Fig. 4. Homologous regions between lima bean lectin decapeptide and other legume lectins.

REFERENCES

Boyd WC, Reguera RM (1949). Hemagglutinating substances for human cells in various plants. J. Immunol 62:333.

Boyd WC, Shapleigh E (1954). Antigenic relations of blood group antigens as suggested by tests with lectins. J Immunol 73:226.

Cunningham BA, Hemperly JJ, Hopp TP, Edelman GM (1979). Favin versus concanavalin A: circularly permuted amino acid sequences. Proc Natl Acad Sci USA 76:3218.

Foriers A, DeNeve R, Kanarek L, Strosberg, AD (1978). Common ancestor for concanavalin A and lentil lectin? Proc Natl Acad Sci USA 75:1136.
Foriers A, DeNeve R, Kanarek L, Strosberg AD (1979). Lectin sequences as a tool for chemotaxonomical classification. Physiol Veg 17:597.
Galbraith W, Goldstein IJ (1970). Phytohemagglutins: A new class of metalloproteins. Isolation, purification, and some properties of the lectin from Phaseolus lunatus. FEBS Letts 9:197.
Galbraith W, Goldstein IJ (1972). Phytohemagglutinin of the lima bean (Phaseolus lunatus). Isolation, characterization, and interaction with type A blood-group substance. Biochemistry 11:3976.
Goldstein IJ (1981). Plant derived lectins. Recent Adv Phytochem 15:25.
Goldstein IJ, Blake DA, Ebisu S, Williams TJ, and Murphy LA (1981). Carbohydrate Binding Studies on the Bandeiraea simplicifolia I isolectins. Lectins which are mono-, di-, tri-, and tetravalent for N-acetyl-D-galactosamine. J Biol Chem 256:3890.
Gould NR, Scheinberg SL (1970a). Isolation and partial characterization of two anti-A hemagglutinins from P. lunatus. Arch Biochem Biophys 137:1.
Gould NR, Scheinberg SL (1970b). Involvement of sulfhydryl groups in the activity of anti-A hemagglutinins of Phaseolus lunatus. Arch Biochem Biophys 141:607.
Hopp TP, Hemperly JJ, Cunningham BA (1982). Amino acid sequence and variant forms of favin, a lectin from Vicia faba. J Biol Chem 257:4473.
Lamb JE, Goldstein IJ (1983). A structural comparison of the A and B subunits of Griffonia simplicifolia I isolectins. Arch Biochem Biophys in press.
Lamb JE, Shibata S, Goldstein IJ (1983). Purification and characterization of Griffonia simplicifolia leaf lectins. Plant Physiol 71:879.
Murphy LA, Goldstein IJ (1977). Five α-D-galactopyranosyl -binding isolectins from Bandeiraea simplicifolia seeds. J Biol Chem 252:4739.
Murphy LA, Goldstein IJ (1979). Physical-chemical characterization and carbohydrate-binding activity of the A and B subunits of the Bandeiraea simplicifolia I isolectins. Biochemistry 18:4999.
Reeke GN Jr, Becker JW, Cunningham BA, Gunther GR, Wang JL, Edelman GM (1974). Relationship between the

structure and activities of concanavalin A. Ann NY Acad Sci 234:369.

Talbot CF, Etzler ME (1978). Isolation and characterization of a protein from leaves and stems of *Dolichos biflorus* that cross reacts with antibodies to the seed lectin. Biochemistry 17:1474.

Acknowledgement

This work was supported by grant GM 29470 from the National Institutes of Health. We thank Linda Carriere for assistance in preparing this manuscript.

Chemical Taxonomy, Molecular Biology, and Function
of Plant Lectins, pages 31–45

STRUCTURAL SUBCLASSES OF LECTINS FROM LEGUMINOUS PLANTS

Joseph W. Becker, Bruce A. Cunningham, and John J. Hemperly

Department of Developmental and Molecular Biology
The Rockefeller University
New York, NY 10021

Plant lectins have been valuable reagents for studies in cell biology, immunology and biochemistry (Lis and Sharon, 1977). These carbohydrate-binding proteins have served as probes for locating and measuring the dynamic properties of cell surface molecules, are among the most effective agents for stimulating mitosis in lymphocytes, and are excellent affinity reagents for isolating glycoproteins (Brown and Hunt, 1978). The function of lectins in plants, however, is unknown, primarily because there have been few unifying principles, other than carbohydrate binding, around which functional assays could be designed.

Over the last seven years, analyses of the structures of lectins have revealed similarities that provide a framework for considering the potential function of the lectins found in leguminous plants. Despite differences in many other properties of these molecules, they are very homologous in amino acid sequence, and are related to each other in unusual ways. The combined data define an entire class of carbohydrate-binding proteins that probably have a common biological function. Moreover, unusual relationships among these molecules should provide an added stimulus to the already burgeoning field of plant molecular genetics.

Lectins belonging to this class are identified as those plant proteins or glycoproteins having significant homology in their amino acid sequences with that of the jack bean lectin concanavalin A (Con A). The prominent position of Con A as a frame of reference is historical in that Con A was the first lectin for which a complete amino acid sequence was described (Cunningham et al., 1975), and it is to date the only lectin of this class to have its three-dimensional structure defined (Becker et al, 1975; Reeke, et al., 1975; Hardman and Ainsworth, 1972). Con A, however, is distinctly different from all of the other lectins identified so far in this class.

The carbohydrate-binding proteins or glycoproteins from leguminous plants can be sorted into three subclasses (Figure 1), the first of which contains only Con A, a protein composed of four identical subunits each containing 237 amino acid residues. The second subclass includes the lectin favin from fava beans (Hemperly et al., 1979), the lectins from lentils and peas (Foriers et al., 1977), and the glucose/-mannose-binding lectins from *Vicia sativa* (Gebauer et al., 1979), and *Vicia cracca* (Baumann et al., 1982). Each of these proteins or glycoproteins is composed of two α chains (M_r = 5000) and two β chains (M_r = 20,000), which in sum are equivalent to one Con A monomer and are presumably arranged as a molecule equivalent to a Con A dimer $(\alpha\beta)_2$. The complete amino acid sequences of both favin (Cunningham et al., 1979; Hopp et al., 1982) and the lentil lectin (Foriers et al., 1981) have been reported. The α chains of these lectins are very homologous in amino acid sequence to each other and to residues 70-119 of Con A. The β chains are homologous to each other and, quite surprisingly, to the rest of the Con A polypeptide; obviously this requires an unusual relationship between these β chains and Con A as shown in Figure 2. The β chains begin at a region corresponding to Con A residue 120 and proceed to the equivalent of the carboxyl terminus of Con A. The homology then continues without interruption through the amino terminus of Con A so that the β chains end near the region in Con A where the favin and lentil lectin α chains begin.

This unusual circular permutation in amino acid sequence with respect to Con A is also apparent in the third lectin subclass, which includes the lectins from soybeans (SBA), peanuts, red kidney beans, and the GalNAc-binding lectins from *Vicia cracca* and *Dolichos biflorous*. At first glance, lectins in this subclass resemble Con A closely in that they are composed of polypeptide chains similar in size to the Con A monomer (see Figure 1), and they are usually tetrameric. Their sequences (Miller et al, 1975; Foriers et al., 1977; Baumann et al., 1979; Etzler et al., 1977), however, begin at positions corresponding to the β chain of subclass 2

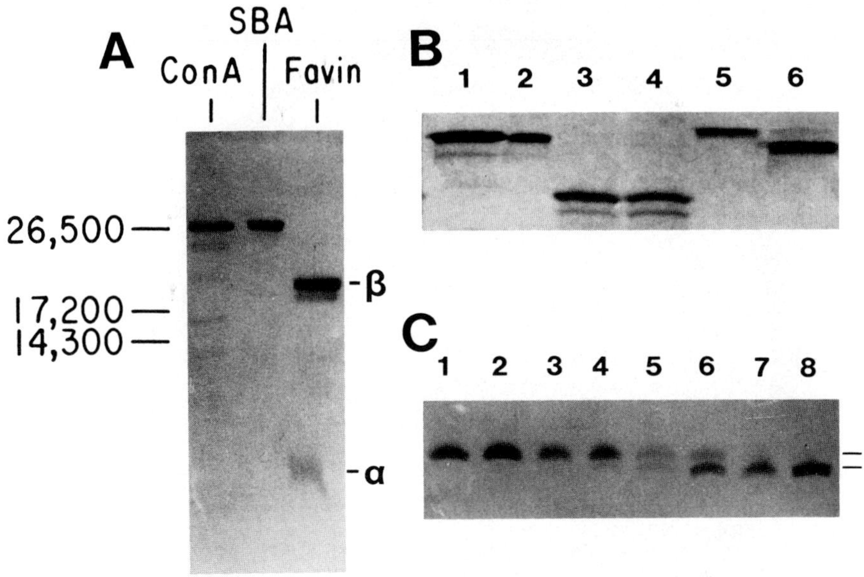

Figure 1. Polyacrylamide gel electrophoresis in SDS of representatives of the three subclasses of leguminous plant lectins. Panel A: Con A (subclass 1), soybean agglutinin (SBA; subclass 3) and favin (subclass 2). The mobilities of standards (M_r=26,000; M_r=17,200; M_r=14,300) and of the α and β chains of favin are indicated. The β' chain of favin migrates directly below the β chain. Panel B: Lane 1, Con A; 2, Con A treated with endoglycosidase F; 3, favin; 4, favin treated as in lane 2; 5, soybean agglutinin; 6, soybean agglutinin treated as in lane 2. Panel C: Lane 1, untreated soybean agglutinin; lanes 2-8, serial 1 to 3 dilutions of endoglycosidase F with the lowest amount of enzyme in lane 2 and the highest in lane 8. The mobilities of the only two components visible are indicated on the right.

lectins, and the equivalent of the α chain of the subclass 2 molecules is at the carboxyl terminus of subclass 3 lectins (Hemperly et al., 1982a), i.e., the polypeptide chains are equivalent to one α chain covalently linked to one β chain of the subclass 2 lectins. To date, subclass 3 contains

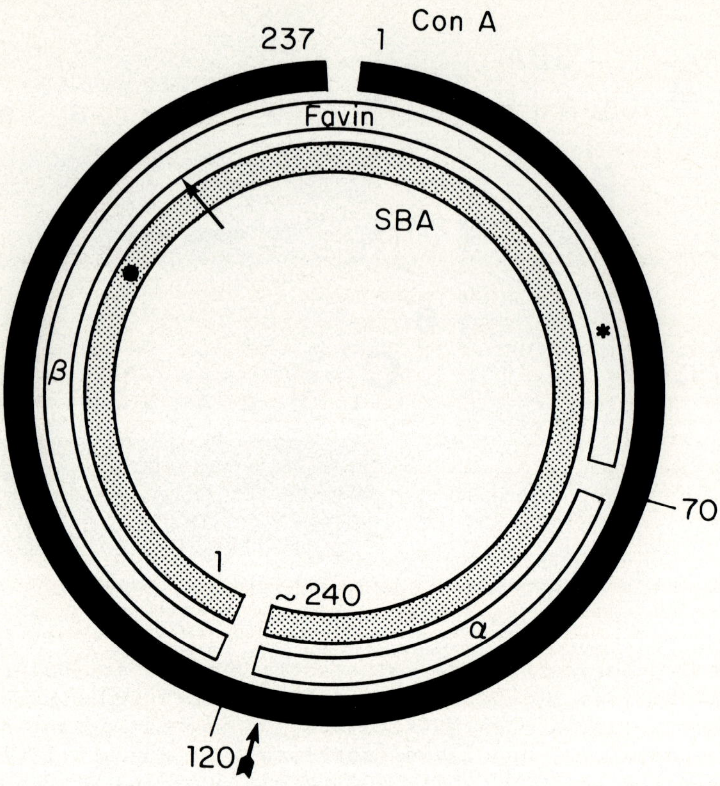

Figure 2. Schematic drawing of favin α and β chains (open bars), soybean agglutinin (stippled bar), and Con A (dark bar) showing the circular permutation that gives maximum homology among the three sequences. The location of the covalently bound carbohydrate of favin (asterisk) and of the tentative location of the sugar of SBA (closed circle) are indicated. The sites of naturally occurring peptide cleavage of Con A (heavy arrow) and of favin (light arrow) are shown. (Adapted from Hemperly et al., 1982a and reproduced by permission of Academic Press, Inc.)

the bulk of the lectins described; it is, however, the least defined group with no well-described representative.

Concanavalin A

Con A is present in jack beans in unusually large amounts; it is easily isolated by affinity chromatography on commercially available supports and readily crystallizes. Its biological effects are both diverse and dramatic. From these properties, it is not surprising that it was the lectin for which a complete structure was first determined. The fact that complete structures for other lectins have not rapidly followed reflects the fact that total structure analysis is still an enormous task and that lectins, for various technical reasons (including a general lack of sulfur-containing amino acids), have proved especially difficult.

In terms of its structural features, each Con A monomer can be considered as a complete unit (Figures 3 and 4). In addition to its 237 amino acids, each monomer contains one Mn^{2+} and one Ca^{2+} that are required for saccharide binding. The saccharide-binding site is located on the top surface of each monomer close to the metal atoms (Becker et al., 1976; Hardman and Ainsworth, 1976). The lower portion of the monomer includes two notable features: a hydrophobic pocket that can bind, among other things, plant auxins (Edelman and Wang, 1978); and two strands of β-structure connected by a loop of peptide where spontaneous cleavage can occur without signigicantly affecting the molecule.

The secondary structure is distinguished by its extensive β-structure. Two large pleated sheets make up the bulk of the molecule. One extends through the center of the monomer and gives gross structure to the subunit. The other forms a flat surface at the back of the molecule and is crucial in both monomer-monomer and dimer-dimer interactions.

Con A is specific for glucose and manose residues and binds to a wide variety of cell surface glycoproteins. It has pronounced effects on the mobility of cell surface receptors and is mitogenic for lymphocytes (Edelman, 1976).

Favin vs. Con A

The lectins belonging to the second subclass also bind specifically to glucose and mannose residues, although their detailed specificity differs somewhat from that of Con A (Allen et al., 1976). They share many biological activities

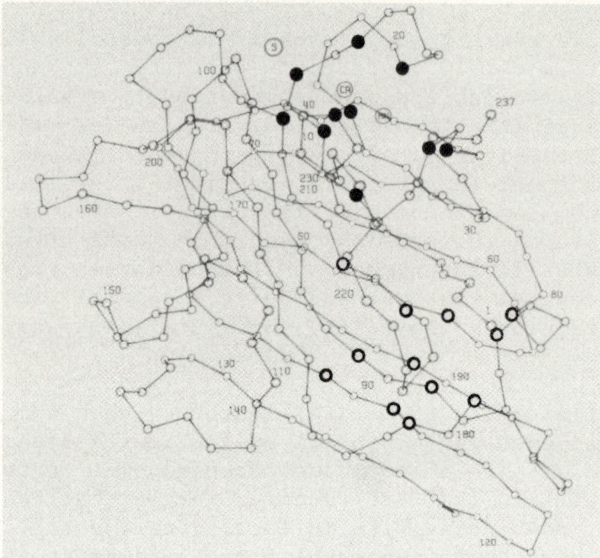

Figure 3. Three-dimensional structure of the Con A protomer. The polypeptide chain is represented by circles at alpha-carbon positions connected by lines. The sites of Mn^{2+}, Ca^{2+}, and specific carbohydrate binding are indicated by MN, CA, and S, respectively. The metal ion ligands are indicated by filled circles and the residues of the hydrophobic cavity by open circles.

with Con A, but appear to bind less strongly to glycoproteins than Con A; for this reason lentil lectin supports are generally preferred over Con A supports for affinity purification of glycoproteins.

While the high-resolution three-dimensional structure of the lectins from this group has not yet been determined, some predictions about the three-dimensional structures seem relatively safe. To illustrate these similarities, we will use our data on favin, but the structure of the lentil lectin (and probably the pea lectin) is so similar that most conclusions will undoubtedly apply to all members of this subclass.

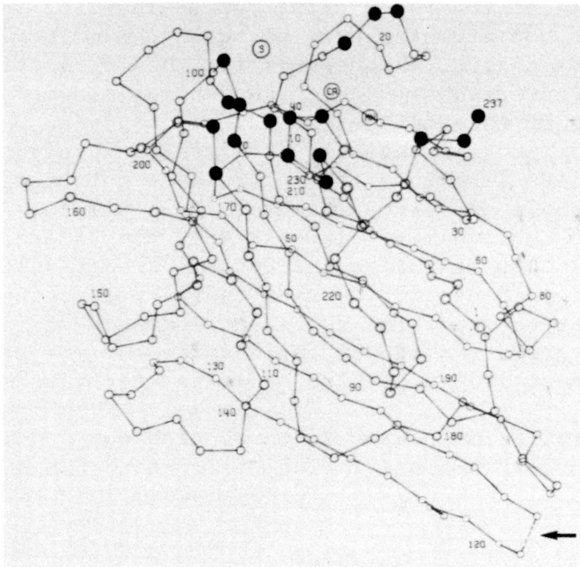

Figure 4. Three-dimensional structure of the Con A protomer showing the residues surrounding the specific carbohydrate-binding site (filled circles) and the site of the naturally-occurring peptide cleavage between residues 118 and 119 (arrow).

When aligned as shown in Figure 2, the amino acid sequences of a Con A monomer and a favin unit (α plus β chain) are very similar. The similarities are particularly striking in the regions of Con A that stabilize its secondary and tertiary structures, suggesting that favin and Con A will resemble each other in three-dimensional as well as primary structures. This notion is also supported by circular dichroism measurements of Con A and favin in solution (Y. Leutzinger, John J. Hemperly, and Bruce A. Cunningham, unpublished observations). Sixty-six percent of the side chains in the interior of the Con A protomer, 56% of the glycine residues, and 63% of the residues in the front β-structure are identical in favin. The back β-structure is much less highly conserved, but the differences in the sequences are concentrated among the residues of this sheet that project out from the back of the Con A molecule. The interior side chains, which stabilize the Con A structure, are mostly conserved. Because the monomer-monomer interactions of the Con A dimer depend mostly on main-chain and non-specific interactions, differences in side chain residues are less likely to influence these interactions, and it is reasonable to assume that the favin molecule will resemble the Con A dimer. In contrast, the dimer-dimer interactions that stabilize the Con A tetramer depend on specific interactions of the side chains projecting out of the back β-structure. The decreased homology of favin in this region could provide an explanation for why favin exists as a species equivalent to Con A dimers, not tetramers.

The detailed molecular features of a Con A monomer and a favin unit are very similar (Table 1). Both Con A and favin

Table 1. Similarities Between Con A and Favin

Structural Feature		Number of Residues in Con A	Identical Residues in Favin
Metal Binding	Mn	4 side chains	4
		2 *via* H_2O	0
	Ca	3 side chains	3
		1 carbonyl (Tyr)	0 (Phe)
		2 *via* H_2O	2
Carbohydrate Binding		18	6
Hydrophobic Cavity		12	8 (4)
Front β-Structure		57	36
Back β-Structure		64	22
Monomer-Monomer		21 non-β	7
		3 β	0
Dimer-Dimer		27	9

contain Mn^{2+} and Ca^{2+}. Of the residues whose side chains are ligands for the metals in Con A, all are identical in favin; of those that contribute carboxyl groups or are secondary ligands (via water molecules), however, some are different. Similarly, the residues that make up the hydrophobic cavity in Con A are largely conserved in favin and of those that are not identical, all are hydrophobic in both lectins. The carbohydrate binding site appears less similar, but this may be deceiving: the exact site has not been defined in Con A and many residues implicated may not be directly involved in sugar binding. In addition, Con A and favin are known to have some differences in their sugar specificities (Allen et al., 1976).

Both Con A and favin, and probably most other lectins, as isolated from beans include some fragmented molecules (Wang et al., 1971; Hopp et al., 1982; Lotan et al., 1975b). Most of these cleavages occur in highly exposed regions of the molecules and appear to have little effect on the structure; even single crystals of these lectins include fragmented molecules. Fragmentation does influence stability in that molecules containing cleaved chains are preferentially denatured (Cunningham et al., 1972). A chief site of such cleavage in Con A is the extended region near the site where the β chain of favin begins. The favin β chain has a primary internal cleavage site at residue 76. The corresponding residue (203) of Con A is on another extended loop near the carbohydrate binding site. Some molecules of favin also include a smaller form of the β chain (β') that lacks carbohydrate and may also be shortened at the carboxyl terminus.

A major chemical difference between Con A and favin is that favin is a glycoprotein. The sugar moiety is variable in structure from molecule to molecule and is small relative to those found in most other glycoproteins. It is located on Asn 168 in the sequence Asn-Ala-Thr (Cunningham et al., 1979). The corresponding sequence in Con A (Asn-Ser-Val) lacks the appropriate specificity for the enzymes known to catalyze sugar attachment to glycoproteins (Marshall, 1972).

As determined so far, the lectins of the same subclass as favin (lentil and pea lectins, among others) are virtually the same as favin in terms of their chemical features and similarities with Con A. The β chain of the lentil lectin is shorter than that of favin and lacks carbohydrate (i.e., it is similar in M_r to the favin β' chain) but the sequence homologies with favin and Con A are extensive. Lentil lectin and favin (and probably pea lectin) are more closely related to each other than either is to Con A. Similarly, the glucose-binding lectins from *Vicia sativa* and *Vicia cracca* appear more similar to favin, pea lectin, and lentil lectin than to Con A (Gebauer *et al*., 1979; Baumann *et al*., 1982).

Soybean Agglutinin

The third subclass of the lectins from leguminous plants has the largest number of representatives, but none has yet been characterized to the same extent as Con A, favin, and the lentil lectin. The lectins in this subclass also have different sugar specificities (primarily galactose/-galactosamine) and might eventually prove to be more dissimilar to Con A than lectins in subclass 2.

Among the lectins in this subclass, we have focused our attention on the soybean agglutinin (SBA). Like most members of this subclass, it appears to contain four polypeptides comparable in size to Con A (Figure 1), but these polypeptides start with a sequence homologous to the β chain of favin. Our preliminary amino acid sequence studies of SBA suggest that it is circularly permuted with regard to Con A (Figure 2) and includes a region comparable to the favin chain at its carboxyl terminus (Hemperly *et al*., 1982a).

SBA is a glycoprotein, but has a larger sugar moiety than favin (Dorland *et al*., 1981). The sugar is mannose rich and is readily cleaved by endoglycosidase F (Elder and Alexander, 1982). Digestion with decreasing amounts of endoglycosidase F give no indication of intermediate species (Figure 1) supporting the notion that there is only one such sugar moiety. We cannot exclude the possibility that SBA has a small additional sugar moiety comparable to that found in favin. Endoglycosidase digestion followed by SDS-PAGE does not change the migration of the favin β chain (Figure 1), and although the sequence in SBA comparable to the region where the carbohydrate of favin in attached contains an Asp where favin has the required Asn, it is possible that the SBA peptides isolated for sequence determination represent a small population of molecules that had lost their sugar moieties.

The major sugar moiety of SBA, however, is not located in the same position as the sugar group in favin; in fact, it is considerably removed in the linear sequence (see Figure 2). Our recent data place it near position 74 of the SBA molecule which is comparable to residue 196 of Con A, at the edge of the tetramer. If SBA resembles Con A in three-dimensional structure, the carbohydrate can project out of the back β-structure and into solution, not interfering with tetramer formation.

Lectins of the third subclass have differing biological properties than those in the first two, probably because of their different sugar specificities. They also appear to include a number of isolectins and, as a result, hybrid species of closely related forms of these molecules appear to exist (Miller *et al.*, 1975; Lotan *et al.*, 1975a; Etzler *et al.*, 1977). This phenomenon is probably a major reason why members of this group have proved difficult to characterize. Much of this difficulty may be circumvented by gene cloning techniques which should provide amino acid sequences that would otherwise be difficult to obtain from mixtures of closely related proteins.

In vitro synthesis of favin

Despite the role of Con A as a paradigm for lectin structure, *in vitro* synthesis studies of favin and the combined sequence data on all the lectins suggest that Con A is the most unusual of this class of molecules. Synthesis of favin *in vitro* gives a single polypeptide precursor (Figure 5) comparable in size to the Con A monomer (Hemperly *et al.*, 1982b). Immunological, peptide mapping and radiochemical sequence data indicate that this product includes both the β chain sequence and the α chain sequence. Translation of *Vicia faba* mRNA in the presence of $[^3H]$-isoleucine, $[^3H]$-leucine, $[^3H]$-phenylalanine, and $[^{35}S]$-methionine yields a product with the sequence:

```
 1                                   10
 x - x -Ile- x - x - x -Met-Leu- x -Phe- x - x -Ile-Phe-Leu
                 20                                      30
 x -Ile-Leu-Leu- x - x -Ile-Leu-Phe-Phe- x - x - x - x - x
                                     40
 x - x -Ile- x - x -Phe- x -Ile- x - x -Phe...
```

The isoleucines at positions 33 and 38 and the phenylalanines

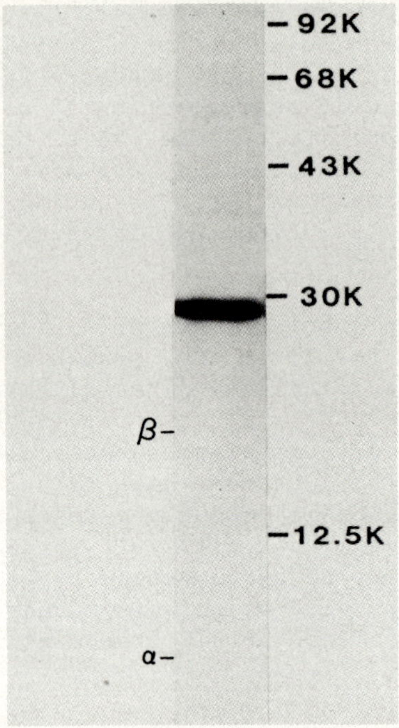

Figure 5. In vitro translation of Vicia faba bean mRNA. Translation was carried out in the presence of [^{3}H]-leucine. Material immunoprecipitated with anti-favin antiserum was subjected to polyacrylamide gel electrophoresis in SDS and visualized by fluorography. The mobility of molecular weight standards are indicated on the right and mobilities of the α and β chains of favin are indicated on the left.

at positions 36 and 41 are consistent with the in vitro product consisting of a hydrophobic 29-residue signal sequence that is followed by the sequence of the β chain i.e., the precursor has the structure: signal peptide: β chain: α chain.

The signal peptide can be removed and the peptide glycosylated by dog pancreas membranes, presumably mimicking normal processing. There is, however, some apparent aberrant processing in the in vitro system (Hemperly et al., 1982b).

Attempts to duplicate the cleavage of the in vitro product to β and α chains have so far failed. Most likely, isolation of the enzymes involved will require more extensive work with immature beans. Our studies on the translation of Con A mRNA and other studies (Vodkin, 1981) suggest that maximal lectin mRNA is present in immature seeds. The polypeptide-processing enzymes are probably also at optimal concentrations in immature seeds.

The evidence that favin is made as a molecule beginning with the β chain indicates that at the genetic level, the subclass 2 lectins resemble closely the lectins in subclass 3. This result suggests that a scheme of the type illustrated in Figure 6 could account for the origins of the

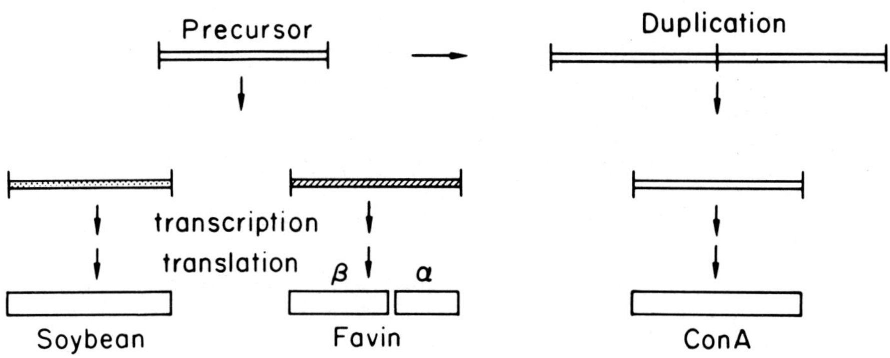

Figure 6. Possible origins of the genes for lectins representing the three subclasses.

genes for these molecules. The scheme is only an example; mechanisms other than gene duplication could give rise to the circular permutation and there is no evidence as yet to indicate the point at which the various genes diverged. In addition, the isolectins found in subclass 3 suggest that gene duplication may be much more prevalent than indicated by this simple diagram. Nevertheless, Figure 6 does give a framework for considering the various possibilities and indicates what now appears to be the appropriate place of Con A with relation to the other lectins.

ACKNOWLEDGEMENTS

We are grateful to Dr. John H. Elder of the Research Institute of Scripps Clinic for the sample of endoglycosidase F. The authors' work described here was supported by USPHS Grant AI-11378.

REFERENCES

Allen, AK, Desai, NN, and Neuberger, A (1976). Biochem J 155:127.
Baumann, CM, Rüdiger, H, and Strosburg, AD (1979). FEBS Lett 102:216.
Baumann, CM, Strosberg, AD, and Rüdiger, H (1982). Eur J Biochem 122:105.
Becker, JW, Reeke, GN Jr, Wang, JL, Cunningham, BA, and Edelman, GM (1975). J Biol Chem 250:1503.
Becker, JW, Reeke, GN Jr, Cunningham, BA, and Edelman, GM. Nature (London) (1976). 259:406.
Brown, JC and Hunt, RC (1978). Int Rev Cytol 52:277.
Cunningham, BA, Wang, JL, Pflumn, MN, and Edelman, GM (1972). Biochemistry 11:3233.
Cunningham, BA, Wang, JL, Waxdal, MJ, and Edelman, GM (1975). J Biol Chem 250:1503.
Cunningham, BA, Hemperly, JJ, Hopp, TP, and Edelman, GM (1979). Proc Natl Acad Sci USA 76:3218.
Dorland, L, van Halbeck, H, Vliegenhart, JFG, Lis, H, and Sharon, N (1981). J Biol Chem 256:7708.
Edelman, GM (1976). Science 192:218.
Edelman, GM, and Wang, JL (1978). J Biol Chem 253:3016.
Elder, JH, and Alexander, S (1982). Proc Natl Acad Sci USA 79:4540.
Etzler, ME, Talbot, CF, and Ziaya, PR (1977). FEBS Lett 82:39.
Foriers, A, Wuilmart, C, Sharon, N, and Strosberg, AD (1977). Biochem Biophys Res Commun 75:980.
Foriers, A, Lebrun, E, Van Rapensbusch, R, de Neve, R, and Strosberg, AD (1981). J Biol Chem 256:5550.
Gebauer, G, Schiltz, E, Schimpl, A, and Rüdiger, H (1979). Hoppe-Seyler's Z Physiol Chem 360:1727.
Gebauer, G, Schiltz, E, and Rüdiger, H (1981). Eur J Biochem 113:319.
Hardman, KD, and Ainsworth, CF (1972). Biochemistry 11:4910.
Hardman, KD, and Ainsworth, CF (1976). Biochemistry 15:1120.

Hemperly, JJ, Hopp, TP, Becker, JW, and Cunningham, BA (1979). J Biol Chem 254:6803.
Hemperly, JJ, Becker, JW, and Cunningham, BA (1982a). In Bradshaw, RA, Hill, R, Tang, J, Liang, C, Tsao, T, and Tson, C (eds): Proteins in Biology and Medicine, New York: Academic Press, p 395.
Hemperly, JJ, Mostov, KE, and Cunningham, BA (1982b). J Biol Chem 257:7903.
Hopp, TP, Hemperly, JJ, and Cunningham, BA (1982). J Biol Chem 257:4473.
Lis, H, and Sharon, N (1977). In Sela, M (ed): The Antigens, New York: Academic Press, p 429.
Lotan, R, Cacan, R, Cacan, M, Debray, H, Carter, WG, and Sharon, N (1975a). FEBS Lett 57:100.
Lotan, R, Lis, H, and Sharon, N (1975b). Biochem Biophys Res Commun 62:144.
Miller, BJ, Hsu, R, Heinrikson, R, and Yachin, S (1975). Proc Natl Acad Sci USA 72:1388.
Marshall, RD (1972) Annu Rev Biochem 41:673.
Reeke, GN Jr, Becker, JW, and Edelman, GM (1975). J Biol Chem 250:1513.
Vodkin, L (1981). Plant Physiol 68:766.
Wang, JL, Cunningham, BA, and Edelman, GM (1971). Proc Natl Acad Sci USA 68: 1130.

Chemical Taxonomy, Molecular Biology, and Function
of Plant Lectins, pages 47–61

STRUCTURAL PROPERTIES OF LEGUME LECTINS

Leland M. Shannon

Department of Biochemistry
University of California
Riverside, California 92521

A large number of legume species have been shown to possess hemagglutinins, and many of these proteins have been purified and well characterized. It has been very difficult to develop meaningful generalities about hemagglutinins primarily because their physiological function(s) is totally unknown and the information required to establish the relationships between different hemagglutinins has not been available. It would be useful to know how many physiologically or evolutionarily distinct classes of protein the hemagglutinins represent, to which class each of the previously characterized hemagglutinins belong, and what (if any) relationship exists between each class.

Amino acid sequence analysis is required for unambiguous evidence of structural relationships among proteins. When proteins are found to be structurally similar, there is a strong probability of both an evolutionary and functional relationship. Determination of the complete amino acid sequence of any protein is difficult and time consuming. Consequently, the complete amino acid sequence has been determined only in lectins from jack bean, Becker *et al*. (1976), soybean, Hemperly *et al*. (1983), *Lens*, Foriers (1981), and *Vicia faba*, Cunningham *et al*. (1979). The amino acid sequence of the amino terminus, however, has been determined in lectins from pea, Foriers (1977), *Vicia sativa*, Gebauer *et al*. (1981), *Vicia cracca*, Baumann *et al*. (1979), *Dolichos*, Etzler *et al*. (1977) kidney bean, Miller *et al*. (1973), peanut, *Crotalaria juncea*, and *Lotus tetragonolobus*, Foriers (1979).

The striking feature of these primary structure studies is the extensive conservation in amino acid sequence among the legume lectins. Details of the amino acid sequence homology are described elsewhere in this book by Dr. A. D. Strosberg and Dr. B. Cunningham. The homologies, however, do not extend to non-legume lectins such as wheat germ agglutinin, Wright (1977), *Ricinus communis*, Cawley *et al*. (1978) and the horseshoe crab *Limulus polyphemus*, Kaplan *et al*. (1977).

IMMUNOCHEMICAL STUDIES

Immunochemistry provides a different but equally powerful method for determining relationships among proteins. Antibodies raised against a given lectin will react with a different lectin only when the two proteins share a common determinant(s). Different proteins can be functionally similar and yet antigenically totally unrelated. However, when proteins are found to be antigenically similar, there is a strong probability of both an evolutionary and a functional relationship. The following experiments were designed to assess the immunological relatedness among legume lectins.

Lectins were screened by Ouchterlony double diffusion against normal rabbit sera and antisera prepared against pure lectins from six legume species. The tests were done on agar plates containing sugars to prevent spurious lectin-glycoprotein interactions.

Figure 1 shows the Ouchterlony double diffusion pattern when Sophora antiserum is challenged with Sophora lectin and Bauhina lectin.

In all cases where a cross-reaction was observed, a single "spur" was present indicating that the cross-reacting lectin was immunologically related, but not identical to the homologous lectin. The results of these tests are summarized in Table 1. Note that 11 of the 12 galactose-specific lectins tested crossreacted with one or more of the antisera raised against a different lectin.

Since lectins are multivalent carbohydrate binding proteins, it was conceivable that the lectins could bind to a carbohydrate component in the antiserum, thereby causing

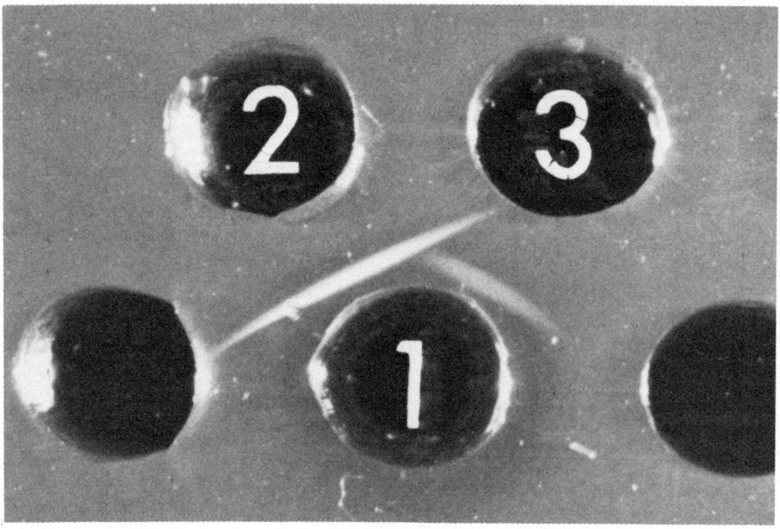

Figure 1. Ouchterlony double diffusion pattern of Sophora antiserum (10 µl in well 1) with Sophora lectin (10 µgr, well 2) and Bauhina lectin (10 µg, well 3).

precipitin bands similar to those obtained in antibody reactions. This possibility was ruled out in the following manner. First, carbohydrate binding was inhibited by saturating the lectin binding sites with sugars. Second, the lectins were tested with purified IgG rather than whole antiserum. These precautions failed to diminish the ability of the antisera to form precipitin bands with the lectins.

As a further test of the specificity of the cross-reactions, we examined the ability of the antisera to inhibit lectin-induced hemagglutination. The hemagglutinating activity of every lectin which cross-reacts with an antiserum, is also specifically inhibited by that serum. The inhibitory potency of each serum was for the most part closely related to the relative strength of the cross reactions seen by Ouchterlony double diffusion. In no case where we could make the test did we observe a cross-reaction between serum and lectin without also seeing inhibition of lectin activity by that serum.

Table 1. Immunological cross-reactions among galactose specific legume lectins

Lectin	Antisera					
	Griffonia simplicifolia	Bauhinia purpurea alba	Glycine max	Phaseolus lunatus	Sophora japonica	Vigna radiata
1. Griffonia simplicifolia (Gal)[a]	⊕	+	-	-	+	+
2. Bauhinia purpurea alba (GalNAc)	+	⊕	-	+	+	+
3. Dolichos biflorus (GalNAc)	-	+	+	+	+	+
4. Glycine max (GalNAc)	-	-	⊕	-	-	-
5. Lotus tetragonolobus (fucose)	+	+	-	-	+	+
6. Phaseolus lunatus (GalNAc)	+	-	+	⊕	-	-
7. Phaseolus vulgaris (Gal)	+	+	-	-	-	-
8. Ricinus communis (Gal)	+	+	-	+	+	+
9. Sophora japonica (GalNAc)	+	+	-	-	⊕	+
10. Ulex europeus (fucose)	+	+	-	-	+	+
11. Vigna radiata (Gal)	+	+	-	-	+	⊕
12. Wisteria floribunda (GalNAc)	-	+	-	-	+	-
Total positive CRM	8	9	2	3	8	7

a. Sugar specificity; ⊕, homologous reaction; +, precipitin band observed; -, no precipitin band observed. From Hankins, Kindinger and Shannon (1979).

All of the legume lectins that have been characterized to date can be placed into two main groups with respect to their carbohydrate-binding specificities: (a) those specific for galactose (and its derivatives such as fucose and GalNAc); and (b) those specific for glucose or mannose (and their derivatives). The overwhelming majority of the legume lectins that have been studied fall into the galactose-specific group.

The results presented here indicate that there is extensive immunological relatedness among the lectins in the galactose group. Of 12 galactose-specific lectins tested, only soybean agglutinin failed to show a cross-reaction with at least one of the sera used. Antisera against soybean lectin did, however, react with two other lectins. We believe that these results coupled with what is already known about these proteins, i.e. they are all lectins, they have similar chemical and physical properties, they are all glycoproteins, they are all from the same plant family, and they are all isolated from seeds, suggest that many of the galactose lectins from legumes are homologs. A similar conclusion has been reached from comparative amino acid sequence analysis and, although the sequence studies involved a very limited number of lectins, the similarities in N-terminal amino acid sequences among several galactose lectins was striking, Etzler et al. (1977) and Foriers (1979).

We have not observed any immunological relationship between the galactose lectins and lectins displaying other specificities, but evidence consistent with an evolutionary relationship has been presented, Foriers (1978). The lack of immunological relatedness between soybean agglutinin and Con A is particularly puzzeling in view of extensive amino acid sequence homologies recently reported by Hemperly et al. (1983).

The evidence now available suggests that most of the well characterized galactose specific legume lectins are members of a specific class of evolutionarily related proteins. Many of these lectins are likely to be homologs.

In view of the above findings and the fact that GalNAc lectins often account for 1% or more of the seed protein, and are in species widely spread within the legume family, we wondered if lectins might be ubiquitious legume proteins. A review of the literature suggested that this possibiility was unlikely, since many legume species, perhaps 50% or more of all species tested, appeared to be totally devoid of hemagglutinin activity of any kind. Furthermore, many of the species which contained hemagglutinins appear to contain proteins with carbohydrate specificities distinct from the GalNAc lectins. To confirm the validity of the earlier literature, a survey of a number of legume species was performed. Indeed, seed extracts of many legume species displayed very little or no hemagglutinin activity, or contained hemagglutinins which were not inhibited by GalNAc. One cannot rule out the presence of a hemagglutinin activity by the limited assay methods used. However, since all of the GalNAc hemagglutinins which have been described are very easily detected in crude extracts by the assay methods used, it is reasonable to conclude that none of these plants contain a detectable quantity of hemagglutinin activity comparable to the GalNAc hemagglutinins.

We then examined extracts of a number of GalNAc hemagglutinin "negative" species to determine if they contained proteins immunologically related to any of the GalNAc hemagglutinins (CRM). The results, given in Table 2, reveal that every plant tested contained CRM to antisera raised against one or more of the the GalNAc hemagglutinins. The antisera used in these studies were from rabbits immunized with very pure hemagglutinin preparations. Although many plants showed CRM with antisera raised against several different lectins, it appeared likely that one major protein was responsible, since adsorption of the extracts by any one of the reactive sera results in a loss of reactivity with all the other sera. This assay is particularly useful because it provides a way to determine the presence of proteins related to lectins, independent of whether they possess hemagglutinin activity.

Most plant hemagglutinins can be purified by affinity chromatography. Sepharose to which N-acetyl galactosamine has been attached provides an excellent adsorbant for the purification of most of the GalNAc specific hemagglutinins. There are now many examples of proteins which have carbohydrate binding properties, but which are not themselves hemagglutinins, that can be purified by carbohydrate affinity chromatography. It is also known that hemagglutinins may sometimes exist in forms, which retain carbohy-

Table 2. Immunological cross reactions between crude seed extracts (which are devoid of hemagglutinin activity) and antisera raised against purified GalNAc specific legume lectins

Seed Extract	Antisera		
	Bauhinia	Dolichos	Sophora
1. Acacia	+	+	+
2. Amorpha	+	-	+
3. Cercis	±	-	+
4. Colutea	+	-	+
5. Delonix	+	±	+
6. Genista	+	-	+
7. Gleditsia	+	-	+
8. Lupinus	+	-	+
9. Mimosa	+	+	+
10. Parkinsonia	±	-	+
11. Poinciana	+	-	+
12. Pueraria	+	+	+
13. Spartium	+	-	+
14. Thermopsis	+	-	+
15. Lespedeza	+	-	+
Total positive CRM	13	3	15

+, precipitin line observed; -, no precipitin line observed; ±, reaction very weak or absent. From Shannon and Hankins (1981).

drate binding properties but are no longer hemagglutinins. This information leads one to question whether or not the CRMs, even though they do not display hemagglutinin activity, can be adsorbed by affinity matricies. Therefore, a number of CRM containing plant extracts were chromatographed on GalNAc sepharose columns with the result that very little, if any of the total protein or CRM was retained by the columns. These findings suggest that the CRMs, which do not display classic hemagglutinin activity, also do not possess the carbohydrate binding properties typical of GalNAc lectins.

The protein responsible for the CRM reaction was purified using standard protein purification procedures.

Antibodies raised to the CRM protein reacted with 6 galactose lectins. These experiments open an exciting area of research. The data indicate that even though many legumes may not contain hemagglutinin activity, they appear to contain proteins immunologically related to lectins. These properties lead us to inquire if the hemagglutinin property, as associated with lectins, is a necessary requirement in the *in vivo* function of these proteins. Perhaps the immunologically related CRM proteins can satifsy this function. Whatever it is!

α-Galactosidases with Hemagglutinin Activity

Working with seed extracts of mung bean, a very potent agglutinin of trypsinized rabbit erythrocytes was observed. The mung bean extract displayed two unique characteristics. Firstly, the hemagglutinin activity was inhibited by xylose and inositol, in addition to galactose, and thus possessed a specificity distinct from previously described hemagglutinins. Secondly, the extracts displayed an interesting property referred to as "clot dissolving activity". Apparently, under the appropriate conditions, this protein is capable of enzymically altering those erythrocyte receptors with which it interacts, resulting in a dissolution of cell aggregates. The disaggregated erythrocytes are permanently altered and are no longer agglutinable by the mung bean extract (although they remain agglutinable by many other legume lectins). Since these observations could be explained by an enzymatic alteration of receptors on the erythrocytes, we examined the mung bean extract for various hydrolytic activities and found a very strong α-galactosidase. During purification we followed both hemagglutin and α-galactosidase activities.

Figure 2 depicts the gel filtration elution profile from a Sephacryl S-200 column, the final purification step. Note that both hemagglutinin and α-galactosidase activities elute at coincident positions. Insert A shows the appearance of SDS gels across the peak. A single band is noted, the intensity of which coincides with A_{280} and activity peaks. Insert B shows the calibration curve of the column. The extrapolated native molecular weight for α-galactosidase-hemagglutin was 160,000. The molecule appeared to be a tetramer composed of a single type of subunit (45,000 MW). All attempts to separate the α-galactosidase from hemag-

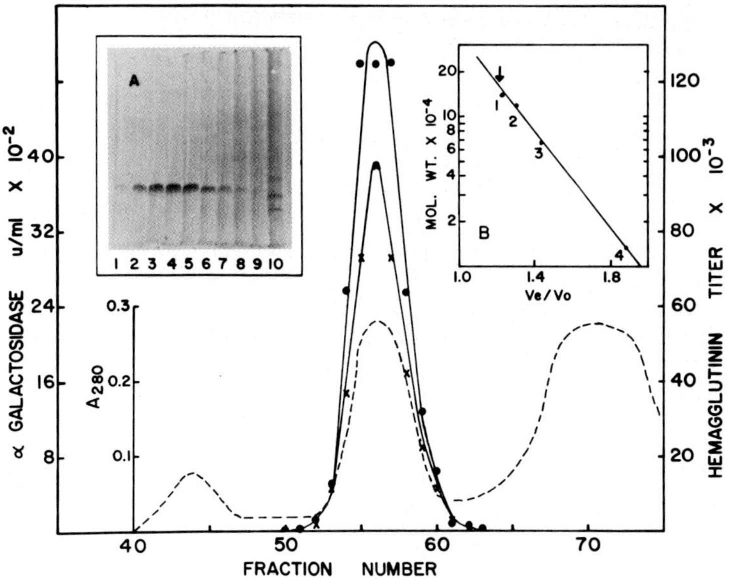

Figure 2. Sephacryl S-200 elution profile. O-O, hemagglutinin; X-X, α-galactosidase; ----, A_{280}. Inset A shows SDS gels of 100 µl portions of fractions across the peak. Inset B shows the calibration curve of the column. From Hankins and Shannon (1978).

glutinin activity were unsuccessful, Hankins and Shannon (1978).

A number of carbohydrates were examined for their ability to inhibit the agglutination and α-galactosidase reactions. The only compounds showing significant inhibitory properties were p-nitrophenyl-α-galactoside [pNPG(α)], galactose, xylose, and inositol. In every case, the concentration of sugar causing 50% inhibition of enzyme activity was very similar to that causing 50% inhibition of hemagglutination, Hankins and Shannon (1978).

In our search for additional α-galactosidase hemagglutinins, we examined seed extracts from 26 genera of legumes, Hankins <u>et al</u>. (1982). We detected clot-dissolving-enzymatic-phytohemagglutinins in 5 additional species. The kinetic and hemagglutinin properties are virtually identical with each other and with the mung bean protein. They possess practically identical K_m values and each extract is inhibited by the same sugars with virtually identical K_i values from plant to plant. The same three sugars inhibited hemagglutinin activity with virtually identical K_i values from plant to plant. The soybean α-galactosidase-hemagglutinin has been purified to apparent homogeniety and appears to be immunologically identical to the mung bean protein, del Campillo and Shannon (1982). The α-galactosidase-hemagglutinin has also been isolated from soybean leaves indicating that this protein is not limited to seed tissues. The α-galactosidase-hemagglutinin in soybean, and lima, are distinct proteins from the well characterized "classic" lectins present in these species. The two groups of proteins can be separated from each other by affinity chromatography or by immunoprecipitation, they have different sugar specificities, and they can exist simultaneously in the same tissue.

The above survey detected clot-dissolving α-galactosidase-hemagglutin activity in only 6 genera. α-Galactosidase activity, however, was present in every legume examined, Hankins et al. (1980). The kinetic properties of hemagglutinin and non-hemagglutinin forms of α-galactosidase are presented in Table 3. The α-galactosidase activities are virtually identical. They possess the same K_m for pNPG(α) and each extract was inhibited by the same sugars with virtually identical K_i from plant to plant. Each extract displayed Michaelis-Menten substrate utilization with competitive inhibition by galactose, xylose and inositol.

We next challenged each of the above extracts with antibody raised against mung bean α-galactosidase-hemagglutinin. Each extract contained materials which cross reacted with the mung bean antibody, i.e., each extract showed CRM - even though the extract did not contain a clot dissolving α-galactosidase-hemagglutinin. This observation led us to inquire if the α-galactosidase activity present in each extract was responsible for the CRM. To answer this, IgG was purified from mung bean antibody and from normal sera. Each IgG was then coupled to Sepharose and extracts from the different legume species, passed over the columns. In each case virtually 100% of the α-galactosi-

Table 3. Kinetic studies of hemagglutinin and non-hemagglutinin forms of α-galactosidase.[a]

	K_m	K_i (mM)		
	(mM)	GAL	Xyl	Inos
Non-Hemagglutinin α-Galactosidases				
Amorpha	1.2	0.5	2.9	12.6
Griffonia	1.1	0.6	4.4	14.7
Bauhinia	1.0	0.5	3.1	15.8
Caragana	0.8	0.6	3.8	9.5
Cercis	1.8	0.5	3.4	9.0
Colutea	0.4	0.4	3.2	17.3
Conavalia	1.2	0.2	1.7	9.5
Cytisus	1.0	0.3	1.8	6.1
Dolichos	1.2	0.5	3.6	--
Genista	0.6	0.3	1.6	11.0
Laburnum	1.4	0.5	2.7	14.7
Lathyrus	1.3	0.6	3.5	8.7
Lens	1.2	1.2	3.3	6.9
Lespedeza	0.5	0.3	2.7	11.7
Mimosa	0.7	0.4	2.2	11.6
Phaseolus	0.7	0.5	3.1	12.8
Sophora	0.6	0.3	2.2	9.0
Spartium	1.7	0.4	3.5	11.4
Ulex	0.7	0.4	1.7	10.0
Wisteria	0.7	0.6	3.6	17.8
Hemagglutinin α-Galactosidases				
Soybean	0.2	1.1	3.8	12.9
Lima bean	0.4	0.3	2.0	15.0
Lupine	0.2	0.3	1.5	7.5
Thermopsis	0.4	0.4	1.6	8.8
Pueraria	0.4	0.9	2.9	8.8
Mung bean	0.3	0.7	5.2	20.0

a. Experiments were performed using pNPG(α) as the substrate. From Hankins, Kindinger and Shannon (1980a, 1980b).

dase activity could be specifically adsorbed from the extract by antisera raised against the mung bean protein.

No specific binding was seen for β-galactosidase, β-glucosidase, or α-mannosidase. These immunoprecipitation studies indicate that the non-hemagglutinin α-galactosidases are immunologically closely related to the mung bean α-galactosidase-hemagglutinin and were at least partially responsible for the CRM. This result, coupled with the similarities in their physical and kinetic properties, suggests that both types of α-galactosidases are members of one specific functional class of protein, i.e., they are homologues, Shannon and Hankins (1981).

Further evidence that agglutinating and non-agglutinating forms of α-galactosidases are homologues was obtained in the following experiments. When red blood cells were treated with non-agglutinating α-galactosidase, washed, and subsequently treated with an agglutinating type of α-galactosidase, the cells failed to agglutinate (although they remained agglutinable by many common legume lectins). Further, when both types of α-galactosidase activity are added simultaneously to red blood cells, 1 unit of non-agglutinating α-galactosidase enzyme activity prevents the agglutination of 10 units of agglutinating α-galactosidase. These observations suggest that agglutinating and non-agglutinating types of α-galactosidase recognize and hydrolyze common receptors on the red blood cell surface.

We suspect red blood cell agglutination by the agglutinating form of the enzyme is a consequence of slow hydrolysis of the receptor, i.e. when the agglutinating type of α-galactosidase binds a cell surface carbohydrate receptor, before it hydrolyzes the receptor, the other binding site(s) may bind receptors on adjacent cells, thus forming a temporary bridge between cells. As a consequence of slow hydrolysis, a sufficient number of cross-linking bridges form to cause agglutination. Non-agglutinating forms of α-galactosidase, on the other hand, may rapidly hydrolyze the receptors and thereby diminish the number of cross-linking bridges such that agglutination is not possible.

Evidence collected to date indicates that the α-galactosidase-hemagglutinins and the "classic" legume hemagglutinins are members of distinct classes of protein: they can be separated from each other; they can exist simultaneously in the same tissue; and they have different sugar specificity. Immunological studies reveal, however, that these two classes of proteins share at least some antigenic determinants. Antisera raised against several "classic" hemagglutinins cross react with mung bean α-galactosidase-hemagglutinin and antisera to the mung bean protein cross react with several "classic" hemagglutins. Thus, while the α-galactosidase-hemagglutinins and the "classic" legume hemagglutinins appear to be members of distinct classes of proteins, immunochemical data suggest they may be evolutionarily related.

SUMMARY

The above data provide immunochemical evidence that lectins isolated from different legume species may be structurally related proteins and are probably homologues. This generalization derives from the great overall similarities in physical, chemical and biological properties of these proteins. More specifically, these proteins appear to be immunologically closely related and possess a high degree of conservation in their primary structures. Acceptance of this generalization implies a common physiological function for these proteins. Hence, in the search for their function, serious consideration must be given to the group as a whole.

A number of legume species were shown to be totally devoid of proteins with hemagglutinin activity. Without exception, however, these plants contained cross reacting material which was immunologically closely related to one or more of the classic lectins. These results suggest that independent of whether a legume seed extract can agglutinate red blood cells, it appears to contain a protein physiologically homologous to the well studied legume lectins. Further, these observations suggest that "hemagglutinin activity" may not be a required manifestation of the "normal" *in vivo* activity of this group of proteins.

All legumes appear to contain a specific α-galactosidase, but only rarely does this enzyme possess hemagglutinin activity. We suspect the hemagglutinin activity associated with the α-galactosidase-hemagglutinin molecule may result from slow hydrolysis of cell surface carbohydrate receptors and may not be important with respect to the general functioning of this class of protein.

It is clear that a great many questions about these interesting proteins remain to be answered. We hope the observations reported here will stimulate additional work and help provide the foundation from which the physiological roles of lectins will ultimately emerge.

LITERATURE CITED

Baumann C, Rüdiger H, Strosberg AD (1979). A comparison the two lectins from Vicia cracca. FEBS Lett 102:216.
Becker JW, Cunningham BA, Reeke GN Jr, Wang JL, Edelman GM (1976). The Molecular Structure of Concanavalin A. In Bittiger H, and Schnebli HP (eds): "Concanavalin A as a Tool," New York: John Wiley and Sons, p 37.
Cawley BD, Hedblom ML, Houston LL (1978). Homology between ricin and Ricinus cummunis agglutinin: Amino terminal sequence analysis and protein synthesis inhibition studies. Arch Biochem Biophys 1290:744.
Cunningham BA, Hemperly JJ, Hopp TP, Edelman GM (1979). Favin versus Concanavalin A: Circularly permuted amino acid sequences. Proc Natl Acad Sci USA 76:3218.
del Campillo E, Shannon LM (1982). An α-galactosidase with hemagglutinin properties from soybean seeds. Plant Physiol 69:628.
Etzler ME, Talbot CF, Ziaya PR (1977). NH_2-terminal sequences of the subunits of Dolichos biflorus lectin. FEBS Lett 82:39.
Foriers A, Lebrun E, Van Rapenbusch R, deNeve R, Strosberg AD (1981). The structure of the lentil (Lens culinaris) lectin. J Biol Chem 256:5550.
Foriers A, Wuilmart C, Sharon N, Strosberg AD (1977). Extensive sequence homologies among lectins from leguminous plants. Biochem Biophys Res Commun 75:980.
Foriers A, deNeve R, Kanarek L, Strosberg AD (1978). Common ancestor for Concanavalin A and lentil lectin. Proc Natl Acad Sci USA 75:1136.
Foriers A (1979). "The Lentil (Lens culinaris) Lectin: Complete Amino Acid Sequence Determination and Structure-Function Relationship with Other Leguminous Seed Lectins," PhD Thesis, Free University of Brussels, p 130.
Gebauer G, Schiltz E, Rüdiger H (1981). The amino-acid sequence of the α subunit of the mitogenic lectin from Vicia sativa. Eur J Biochem 113:319.

Hankins CN, Shannon LM (1978). The physical and enzymatic properties of a phytohemagglutinin from mung beans. J Biol Chem 253:7791.
Hankins CN, Kindinger JI, Shannon LM (1979). Legume lectins. I. Immunological cross-reactions between the enzymic lectin from mung beans and other well characterized legume lectins. Plant Physiol 64:104.
Hankins CN, Kindinger JI, Shannon LM (1980a). Legume α-galactosidases which have hemagglutinin properties. Plant Physiol 65:618.
Hankins CN, Kindinger JI, Shannon LM (1980b). Legume α-galactosidase forms devoid of hemagglutinin activity. Plant Physiol 66:375.
Hemperly JJ, Cunningham BA (1983). Circular permutation of amino acid sequences among legume lectins. TIBS 8:100.
Kaplan R, Li SS-L, Kehoe JM (1977). Molecular characterizatin of limulin, a sialic acid binding lectin from hemolymph of the horseshoe crab, Limulus polyphamus. Biochem 16:4297.
Miller JB, Noyes C, Henrikson R, Kingdon HS, Yachnin S (1973). Phytohemagglutinin mitogenic proteins; Structural evidence for a family of isomitogenic proteins. J Exp Med 138:939.
Shannon LM, Hankins CN (1981). Enzymatic properties of phytohemagglutins. In Loewus FA, Ryan CA (eds): "The Phytochemistry of Cell Recognition and Cell Surface Interactions," New York: Plenum Press, p 92.
Wright CS (1977). The crystal structure of wheat germ agglutinin at 2.2 Å resolution. J Mol Biol 111:439.

Chemical Taxonomy, Molecular Biology, and Function of Plant Lectins, pages 63–70

TOMATO (LYCOPERSICON ESCULENTUM) LECTIN AND SEROLOGICALLY RELATED MOLECULES

David C Kilpatrick

Regional Blood Transfusion Centre,
Royal Infirmary, Edinburgh, U.K.

Solanaceous plants are a source of a number of lectins and lectin-related glycoproteins. Although haemagglutinins were reported in Datura (Eisler & Portheim, 1908; Boyd & Reguera, 1949) and potato (Solanum tuberosum) (Marcusson-Begun, 1926) plants during the first half of the century, it is only in recent years that highly purified forms of Datura stramonium seed (Horejsi & Kocourek 1978; Kilpatrick & Yeoman, 1978; Desai et al, 1981), potato tuber (Allen et al, 1978), potato fruit (Kilpatrick, 1980a) and tomato fruit (Kilpatrick, 1980b; Nachbar et al, 1980) lectins have been isolated and characterised. These purified lectins show many similarities which as a group distinguish them from lectins occurring in other plants (Table 1). Their most noteworthy features include an unusual amino acid composition, with a high hydroxyproline content; a very high carbohydrate content, most of which is arabinose; and a susceptibility to inhibition by oligomers of N-acetylglucosamine, particularly $(GlcNAc)_4$, but not to monomeric N-acetylglucosamine itself. Despite having many characteristics in common, these lectins are unequivocally distinct molecular entities (Table 2). They differ most obviously in molecular size determined by sodium dodecyl sulphate-polyacrylamide gradient gel electrophoresis (and by other methods); in the pattern of inhibition by complex carbohydrates; and in terms of relative agglutinating activity towards erythrocytes from different species. There also appear to be minor differences in amino acid composition, despite an overall similarity. In conclusion, therefore, these lectins are related but not identical molecules.

TABLE 1 GENERAL PROPERTIES OF PURIFIED SOLANACEOUS LECTINS

Agglutinate human erythrocytes; A,B,O non-specific

Readily inhibited by oligomers of GlcNAc, but not by monomeric GlcNAc itself

High carbohydrate content : 40-50% CHO (mainly arabinose)

Distinctive amino acid composition; includes high (14-20% of residues) hydroxyproline content

High thermal stability (70°C; $\geq$ 30 min)

Sensitive to periodate

Insensitive to EDTA; no known divalent cation requirement

Relatively resistant to proteolytic enzymes

Exhibit immunological cross-reactivity

The above conclusion was supported by serological studies using a rabbit antiserum raised to the lectin prepared from Datura stramonium seeds. The anti-(Datura lectin) serum cross-reacted in double diffusion experiments with both tomato fruit and potato fruit/tuber lectins and was able to precipitate all three from solution in the presence of polyethylene glycol (Kilpatrick et al, 1980). Extracts containing haemagglutinating activity from other Solanaceous sources also cross-reacted with the antiserum but were not readily inhibited by it. An unexpected finding was that all the seed lectins examined (except Datura) formed a continuous line of identity when placed in adjacent wells. These seed lectins differed, however, in their susceptibility to inhibition by chitobiose or a mixture of N-acetylglucosamine oligomers, and the pattern of saccharide inhibition was consistant with accepted taxonomic relationships. Seed lectins and "non-seed" lectins formed lines of partial identity with each other, even if they were from the same species. The lectins from tomato fruits, potato fruits and potato tubers formed lines of identity with this antiserum.

An alternative and highly sensitive method of invest-

TABLE 2 DIFFERENCES IN PROPERTIES OF SOLANACEOUS LECTINS

Source of Lectin	Mr (SDS-gradient PAGE)	Inhibition by 1mg/ml thyroglobulin (% control activity)	Agglutinating activity human: sheep cells
Tomato (fruit)	68,000	10	30:1
Potato (fruit)	55,000	20	10:1
Datura stramonium (seed)	30,000	1	1:1

igating antigenic similarity is the technique of indirect immunofluorescence. Thin sections prepared from stems and other tissues of various species were treated first with anti-(*Datura* lectin) immunoglobulin followed by xenogeneic anti-immunoglobulin. By this means, not only were the immunodiffusion cross-reactivities confirmed, but specific staining was also found in representatives of several dicotyledonous families other than the Solanaceae (Jeffree et al, 1982). In all cases, the specific fluorescence was cytoplasmic, apparently associated with membranous structures.

Such studies prompted the following suggestions : (1) the seed lectins of the Solanaceae form an antigenic group distinct from the lectins found in other tissues; (2) these seed lectins (and non-seed lectins) have in common one or more epitopes that have been conserved through evolution, and might therefore be of major functional importance; (3) since the carbohydrate specificity of these lectins has not been conserved, the precise carbohydrate specificity is not essential to their function. This third suggestion appears at variance with the general belief that the function of a lectin must be related to its saccharide specificity, but a simple explanation for this apparent paradox is possible. Complex carbohydrates often contain a number of different carbohydrate structures on the same molecule, any one of which might serve as a receptor site for a given lectin. If the function of a lectin involved binding to a particular glycoprotein, the precise point of attachment need not be important. Two lectins with quite different specificities determined by simple sugar inhibition tests are capable of binding to the same complex carbohydrate (e.g. lectins of different specificities can bind to thyroglobulin). Therefore evolutionary changes resulting in changes in saccharide specificity might be tolerated if the lectin could still bind to its natural substate, even though the binding were mediated by different receptor sugars.

The tomato plant has lectin activity in various tissues, with the highest specific activity found in the pericarp and the juice within it (Kilpatrick,1980b). The tomato seed has an apparently separate but cross-reacting (with anti-*Datura* lectin serum) lectin, which differs in being much less sensitive to inhibition by oligomers of N-acetylglucosamine (Kilpatrick et al, 1980). If the anti-*Datura* lectin antiserum contains two types of antibody, one directed against the seed-common epitope(s) and the other(s) recognising structures on phylogenetically related lectins, then an antiserum to the tomato fruit lectin should provide useful complementary information. The double diffusion reactions should be consistant with the results obtained with the anti-(*Datura* lectin) serum, and should help to clarify the relationship between the seed and fruit lectins of the tomato.

Some of the immunodiffusion reactions of anti-(tomato lectin) serum are shown in Figure 1. The antiserum cross-reacted with the *Datura* seed lectin and with the lectin from potato fruits; these formed lines of identity. It did not, however, form a detectable line of precipitation with seed extracts from either the sweet pepper or the tomato itself, although identical preparations reacted

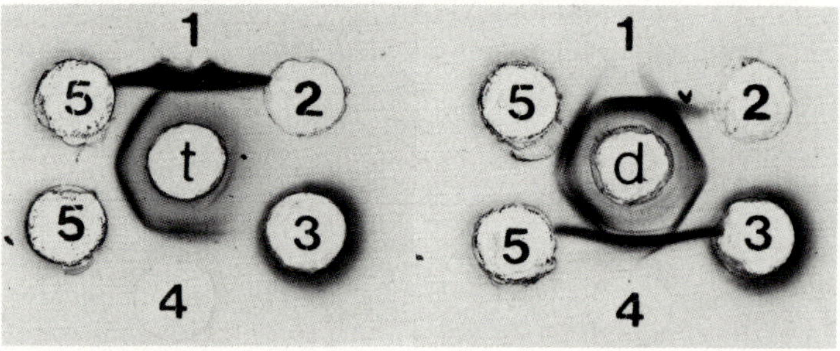

Fig. 1. Double diffusion of anti-(tomato lectin) serum (t) or anti-(*Datura* lectin) serum (d) against tomato lectin (1), tomato seed homogenate (2), sweet pepper seed homogenate (3), *Datura* lectin (4) and potato lectin (5).

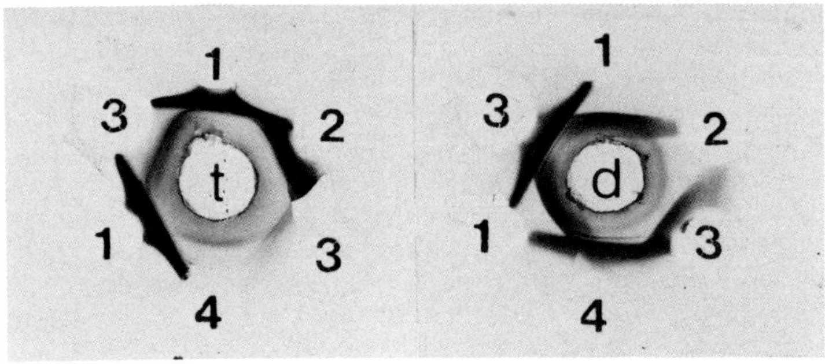

Fig.2. Double diffusion of anti-(tomato lectin) serum (t) or anti-(*Datura* lectin) serum (d) against tomato lectin (1), deglycosylated tomato lectin (2), *Datura* lectin (3) and deglycosylated *Datura* lectin (4).

strongly with the anti-*Datura* lectin serum tested at the same time.

Further investigation into the nature of these serological reactions was made using deglycosylated lectins (Fig. 2). Purified tomato fruit and *Datura* seed lectins were treated with trifluoromethanesulphonic acid according to Edge et al (1981). This treatment produced lectins containing little or no carbohydrate judged by increased mobility during electrophoresis in SDS-containing polyacrylamide gels, and by the absence of staining for carbohydrate with periodic acid-Schiff reagent (Glossman & Neville, 1971) or thymol-H_2SO_4 (Racusen, 1979). Deglycosylated tomato lectin reacted strongly with anti-tomato lectin serum, but not with anti-*Datura* serum; deglycosylated *Datura* lectin reacted with anti-*Datura*, but more weakly with anti-(tomato lectin) serum.

Taken together, these serological reactions can be explained by assuming the existence of at least three domains in the lectin molecules. The first domain, which is without attached carbohydrate, is antigenically unique to each lectin. The second domain, which may contain both glycosylated and non-glycosylated epitopes, is shared by closely related lectins. The third domain, consisting mainly of glycopeptide, is similar even in distantly related molecules. Antibodies to epitopes on the first two domains would account for the cross-reactivity observed between Datura, tomato and potato lectins irrespective of the specific antiserum used, while the cross-reactivity with tissues of distantly related plant species could be due to antibodies directed against the third (glycopeptide) domain. The relative immunogenicity of the three domains may vary with individual rabbits, thereby fully explaining the patterns of reactions. This model is also consistant with the observations of Ashford et al (1982) and Desai et al (1983) using rabbit antisera raised against potato tuber lectin.

The apparent evolutionary conservation of the glycosylated domain implies an important function for the glycosylated moiety, and therefore a comparison of the properties of native and deglycosylated tomato lectins was carried out. No appreciable difference was found in thermal stability, stability to low pH, or susceptibility to proteolysis by Pronase; it seems unlikely, therefore, that the role of carbohydrate is to protect the molecule from degradation or denaturation. The specific activity of the deglycosylated lectin was very much less than that of the native material, possibly suggesting involvement of carbohydrate in the saccharide binding site. When sheep erythrocytes were used instead of human cells, however, the specific activities of native and deglycosylated lectin were found to be similar. Unlike the native lectin which agglutinates human cells (of any blood group) at least thirty-times more readily than sheep cells, the deglycosylated lectin has a human : sheep cell relative agglutinating ratio of only 2:1. This may be an indication that the carbohydrate structure influences saccharide specificity, but both native and deglycosylated forms displayed a quantitatively similar susceptibility to inhibition by thyroglobulin or N-acetylglucosamine oligomers. It is possible, therefore, that the carbohydrate structure is necessary for the maintainance of the conformation necessary to bring about optimal agglutination of human erythrocytes without actually changing the saccharide specificity.

It is a plausible suggestion that the Solanaceous lectins originated from a common ancestor. Some differences between lectins from different species might be simply the result of mutations occurring during evolution, but the existence of antigenically-distinct seed and fruit lectins in the tomato suggests independent evolution probably following gene duplication. The existence of isolectins in, for example, tomato fruits or potato tubers, might also arise from gene duplication and could be the first step towards separate development. The two major isolectins from the tomato have been isolated and compared, but apart from isoelectric point, no difference could be detected in their characteristics.

Since the tomato fruit lectin appears more similar to the potato fruit/tuber lectins than to the lectin in tomato seeds, and the corresponding situation is true of the potato lectins, it is possible that the seed lectins share a common function that is quite different from that of the lectins occurring in fruits. The seed lectins, for example, might be involved with the storage or utilsation of food reserves, while the fruit lectins might confer non-specific protection by neutralising chitin-containing fungal pathogens in an antibody-like manner. The actual function(s) of all of these lectins is, however, unknown; but we should not, perhaps, be looking for a single function common to all lectins, or even a function common to different lectins from a single species.

The tomato lectin is able to agglutinate human lymphocytes and granulocytes as well as erythrocytes. Current research in this laboratory is into the effect of tomato lectin on the former. Nachbar et al (1980) reported an anti-mitogenic action on chicken lymphocytes, and we have found a similar antagonistic effect on PHA or pokeweed mitogen-induced transformation of human peripheral blood lymphocytes. Typically, low doses of tomato lectin reduced tritiated thymidine incorporation to about 75% of the control value. A more potent inhibition is often observed with antigen (e.g. tuberculin) mediated transformation. Such an effect on cell division is obviously compatible with a defence function for the lectin, but like the agglutination of animal cells, this property is probably fortuitous and of no relevance to the physiological role of the lectin.

Acknowledgements

I thank the Chief Scientists Organisation of the Scottish Home and Health Department for financial support and Dr. C.E. Jeffree for help in raising antisera.

References

Allen AK, Desai NN, Neuberger A, Creeth JM (1978). Properties of potato lectin and the nature of its glycoprotein linkages. Biochem J 171:665

Ashford D, Allen AK, Neuberger A (1982). The production and properties of an antiserum to potato (Solanum tuberosum) lectin. Biochem J 201:641

Boyd WC, Reguera RM (1949). Haemagglutinating substances for human cells in various plants. J Immunol 62:333

Desai NN, Allen AK, Neuberger A (1981). Some properties of the lectin from Datura stramonium (thorn-apple) and the nature of its glycoprotein linkages. Biochem J 197:345

Desai NN, Allen AK, Neuberger A (1983). The properties of potato (Solanum tuberosum) lectin after deglycosylation by trifluormethanesulphonic acid. Biochem J 221:273

Edge ASB, Faltynek CR, Hof L, Reichert LE, Weber P (1981). Deglycosylation of glycoproteins by trifluormethanesulphonic acid. Anal Biochem 118:131

Eiser M, Portheim L (1908). Ueber ein hamagglutinin in samen von Datura Z Immunitatsforsch 1:151

Glossman H, Neville DM (1971). Glycoproteins of cell surfaces, A comparative study of three different cell surfaces of the rat. J Biol Chem 246:6339

Horejsi V, Kocourek J (1978). Isolation and characterisation of the lectin from jimson-weed seeds (Datura stramonium). Biochim Biophys Acta 532:92

Jeffree CE, Yeoman MM, Kilpatrick DC (1982). Immunofluorescence studies on plant cells. Int Rev Cytol 80:231

Kilpatrick DC (1980a). Isolation of a lectin from the pericarp of potato (Solanum tuberosum) fruits. Biochem J 191:273

Kilpatrick DC (1980b). Purification and some properties of a lectin from the fruit juice of the tomato (Lycopersicon esculentum). Biochem J 185:269

Kilpatrick DC, Yeoman MM (1978). Purification of the lectin from Datura stramonium. Biochem J 175:1151

Kilpatrick DC, Jeffree CE, Lockhart CM, Yeoman MM (1980). Immunological evidence for structural similarity among lectins from species of the Solanaceae. FEBS Lett 113:129

Marcusson-Begun H (1926). Untersuchungen uber das hamagglutinin der kartoffelknolle. Z Immunitatsforsch 45:49

Nachbar MS, Oppenheim JD, Thomas JO (1980). Lectins in the US diet. Isolation and characterisation of a lectin from the tomato (Lycopersicon esculentum). J Biol Chem 255:2056

Racusen D (1979). Glycoprotein detection in polyacrylamide gel with thymol and sulphuric acid. Anal Biochem 99:474

Chemical Taxonomy, Molecular Biology, and Function of Plant Lectins, pages 71–85

POTATO LECTIN - A GLYCOPROTEIN WITH TWO DOMAINS

Anthony K. Allen, Ph.D.

Department of Biochemistry,
Charing Cross Hospital Medical School,
Hammersmith, London, W6 8RF, England

Potato (*Solanum tuberosum*) tubers were first reported to contain an erythrocyte-agglutinating activity by Gelhorn (1925). This was followed by the work of Marcusson-Begun (1925), who demonstrated that extracts were capable of agglutinating various animal erythrocytes. Later Krüpe (1956) tested the agglutinating activity of extracts of 67 different varieties of potato against human A, B, O blood groups and also against the rabbit, guinea pig, mouse and sheep erythrocytes, showing that potato lectin was a panagglutinin. It has since been shown that its specificity is directed towards oligomers of N-acetylglucosamine and that it is a glycoprotein with two distinct domains, a binding region which is unglycosylated but rich in cystine and a hydroxyproline rich, highly glycosylated region which is not involved in ligand binding. In this paper I will describe some of the work done by ourselves and others which has lead to these conclusions.

PURIFICATION

Early attempts to purify potato lectin were made by Singh *et al.* (1961) by acetone precipitation. However, they described the active extract as being "nucleic acid-like". Marinkovich (1964) partially purified the lectin by acid/alcohol extraction, acetone precipitation and ion exchange chromatography; although claimed to produce a pure lectin, this method was subsequently (Allen, Neuberger 1973) shown to yield a product that was only 10% lectin. The lectin was purified to homogeneity (Allen, Neuberger 1973, 1978) by successive ammonium sulphate precipitation, DEAE-cellulose, CM-cellulose and Sephadex chromatography. The first effective affinity chromatography system for the lectin purification involved the preparation of 4-aminobenzyl 2-acetamido-2-deoxy-1-thio-β-D-glucoside and its coupling to Sepharose (Delmotte *et al.* 1975). This matrix effectively absorbed the lectin from potato tuber extracts and permitted its subsequent elution with acetic acid. The main disadvantage of this technique is that the absorbant required a lengthy synthetic procedure. Subsequently two groups reported that tri-N-acetyl chitotriose coupled to epoxy-activated Sepharose (Masumoto *et al.* 1979; Desai, Allen 1979) was an effective affinity chromatography absorbant for potato lectin which would allow subsequent elution with 0.1-M acetic acid. We have since found that 2 mM-diethylamine is a more effective eluent than acetic acid and we routinely use this reagent for elution (Desai *et al.* 1983). The reported yields have varied from 8 mg/kg (Allen, Neuberger 1973, 1978) to 48 mg/kg (Desai, Allen 1979). As has been shown by the work of Krüpe (1955) there is a great difference (at least 64-fold) in the amount of lectin in different varieties and we have emphasised that it is essential to test as many of the available varieties as possible before embarking on a preparation (Allen, Neuberger 1978).

COMPOSITION

The composition of the lectin is given in Table 1. The values for the amino acids are calculated as molar proportions by assuming that the molecular weight of the monomer is 50,000. These figures vary slightly from one preparation to another, as can be seen from our publications (Allen, Neuberger 1973; Allen *et al.* 1978; Ashford *et al.* 1981; Desai *et al.* 1983). The lectin is a glycoprotein with 50% by weight of carbohydrate, arabinose and galactose are the only sugars. An unusual feature of the lectin is that hydroxyproline is the most abundant of the amino or imino acids. There is also a very high content of cystine and there are no free-SH groups. About 22 of the 29 aspartic and glutamic residues are present as the amides (Ashford *et al.*, 1981), which explains the high pI values of 9.4 to 9.6 which we (Allen *et al.* 1978) observed for isolectins of potato lectin. Various preparations have contained either phenylalanine or histidine but rarely both amino acids in integral quantities. All the preparations that we have analysed have shown the

Table 1. Amino acid and carbohydrate composition of potato lectin expressed as moles/mol of 50,000 mol. wt. monomer.

Hyp	50	Met	1	Amide	22
Asx	12	Ile	4		
Thr	14	Leu	3		
Ser	31	Tyr	8		
Glx	17	Phe	0-1	Ara	172
Pro	17	His	0-1	Gal	10
Gly	30	Lys	9		
Ala	10	Orn	0.3-1		
½Cys	26	Arg	3		
Val	1	Trp	8		
		Total	247		

presence of a basic amino acid which co-chromatographs with ornithine in a number of systems. The presence of ornithine has been confirmed in other laboratories (Muray, Northcote 1978; Masumoto *et al.* 1983). It is most likely that ornithine is produced by the action of a hypothetical protein arginase on arginine residues in the polypeptide chain. It is noteworthy that ornithine has also been reported in arabinogalactan glycoproteins (Jermyn, Yeow 1975).

PHYSICOCHEMICAL CHARACTERISTICS

Studies on the molecular weight of potato lectin have been complicated by the high proportion of carbohydrate in the molecule and possibly by its affinity for the sugar residues of some of the gels which are used for chromatography. Our early studies (Allen, Neuberger 1973) on polyacrylamide gels in SDS and mercaptoethanol gave an apparent molecular weight of 105,000 (±10,000), because of the high carbohydrate content this was recognised to be a considerable overestimate. Analytical ultracentrifugation gave a sedimentation coefficient of 3.37S which gave a minimum molecular weight of 24,000 (assuming f/fo = 1.0). Gel filtration of both the native lectin and the SCMC lectin in 6 M-guanidine HCl gave molecular weights of 46,000. Later studies by J.M. Creeth (Allen *et al.* 1978) gave a sedimentation coefficient of 3.18S in non-denaturing media, but the molecular weight estimates gave indications of an aggregating system. The monomer molecular weight was calculated to be close to 50,000 and the extent of dimer formation to be 30% at 1 mg/ml. The viscosity was very high and shape predictions were either of a very expanded sphere or a prolate ellipsoid (f/fo = 2.4) about 30 times as long as it is wide, or some structure intermediate between these two. Ultracentrifugation in denaturing media gave a molecular weight of about 48,000. The conclusion from these various studies is that potato lectin is probably an elongated molecule made up of monomers of about 50,000 existing in a monomer-dimer association with the monomer predominating.

STUDIES ON THE GLYCOSYLATED REGION OF POTATO LECTIN

Extensive digestion of the SCMC-derivative of potato lectin yielded, apart from individual amino acids and small peptides, a single large glycopeptide (Allen *et al.* 1978) and this was shown to have a molecular weight of 32,000-34,000. The glycopeptide included all the carbohydrate and hydroxyproline of the original glycoprotein but none of the cystine residues and hardly any of the aromatic residues. Since the glycopeptide showed no binding capacity for red cells or oligosaccharides we proposed that potato lectin consisted of two very dissimilar domains; a hydroxyproline-rich glycosylated domain and a cystine-rich binding region which lacked covalently bound saccharides (Table 2).

Attempts were then made to determine the structure of the glycosylated region. We (Allen *et al.* 1978) were able to show that treatment of the glycopeptide with mild acidic conditions (1 M-HCl at 30°C for 10 days) completely removed the arabinose residues but left an almost intact glycopeptide with the galactose residues still attached. From comparative hydrolytic studies with model compounds we concluded that the arabinose was present as a furanoside and the galactose as a pyranoside. It was also concluded from the change in optical rotation of the glycopeptide on removal of the arabinose that the sugar was mostly present as the β-L-arabinofuranoside. By treatment with 5 M-NaOH and subsequent gel filtration hydroxyproline arabinosides were isolated. The action of even mild alkali on the galactose residues was to bring about their release from serine residues by a process of β-elimination; the process was more rapid in the absence of the arabinofuranoside residues and we proposed that a negative charge on the arabinofuranosides inhibited the β-elimination.

Table 2. The composition of the domains of potato lectin.

	Whole lectin	Glycosylated domain	Non-glycosylated domain
Hyp	50	50	0
Asx	12	2	10
Thr	14	2	12
Ser	31	17	14
Glx	17	4	13
Pro	17	4	13
Gly	30	5	25
Ala	10	4	6
½ Cys	26	0	26
Val	1	0	1
Met	1	0	1
Ile	4	1	3
Leu	3	3	0
Tyr	8	0	8
Phe	1	1	0
His	0	0	0
Lys	9	1	8
Orn	1	N.D.	N.D.
Arg	3	1	2
Trp	8	0	8
Total amino acids	246	95	150
Ara	172	172	0
Gal	10	10	0

The substitution of the galactose on the serine residues was shown to be α since it was removed by an α-galactosidase but not by a β-galactosidase.

Further studies (Ashford *et al.* 1982a) led to the complete elucidation of the structure of the carbohydrate moieties. Methylation analysis confirmed that the lectin contained both arabinofuranosides and galactopyranosides. The arabinofuranosides were shown to be present as short unbranched chains with 1,2 and 1,3 linkages and the galactopyranosides as terminal substituents. Alkaline digest of the lectin followed by gel filtration was used to show that all the arabinose was attached to hydroxyproline and that four fractions could be isolated corresponding to the mono-, di-, tri- and tetra-arabinosides, with the latter two accounting for 70% of the total hydroxyproline. Methylation analysis was then used to show that the tri-arabinosides contained only 1,2-linkages between sugars, but that the tetra-arabinoside contained both 1,2 and 1,3 linkages. Direct-insertion mass spectrometry of these compounds using electron impact and chemical ionization was used to determine the sequences of the sugars which were Ara*f*1→2Ara*f*1→2Ara*f*1→Hyp and Ara*f*1→3 Ara*f*1→2Ara*f*1→2Ara*f*1→Hyp. By measurements of optical rotations of the hydroxyprolyl arabinosides and the action of a specific α-arabinosidase we were able to conclude that the trisaccharide contained only β-furanosides but that the tetrasaccharide had an additional terminal α-furanoside. The structures are therefore:-

Trisaccharide: βAra*f*1→2βAra*f*1→2βAra*f*1→Hyp.
Tetrasaccharide: αAra*f*1→3βAra*f*1→2βAra*f*1→2βAra*f*1→Hyp.

The full structures of the saccharides are shown in Fig. 1. It is noteworthy that the structures shown in Fig. 1 are identical with those found in cell wall preparations of tobacco cell walls (Akiyama *et al.* 1980) and presumably with those of hydroxyproline-rich glycopeptides of other higher plants. This means that potato lectin is likely to be a useful soluble model for studies of the biosynthesis of insoluble hydroxyproline-rich glycoproteins in cell walls, as has already been indicated (Owens, Northcote 1981) by preliminary studies.

DEGLYCOSYLATION OF POTATO LECTIN

We have used the trifluoromethanesulphonic acid reagent of Edge *et al.* (1981) to completely deglycosylate potato lectin (Desai *et al.* 1983). Such preparations retained their haemagglutinating activity, had an intact polypeptide chain and had identical amino acid analyses to the native lectin but had no carbohydrate. The deglycosylated lectin cross reacted with one component of an anti-potato lectin (rabbit) antiserum (Ashford *et al.* 1982b) which recognises the polypeptide chain but not against a component which recognises the carbohydrate (see Fig. 2).

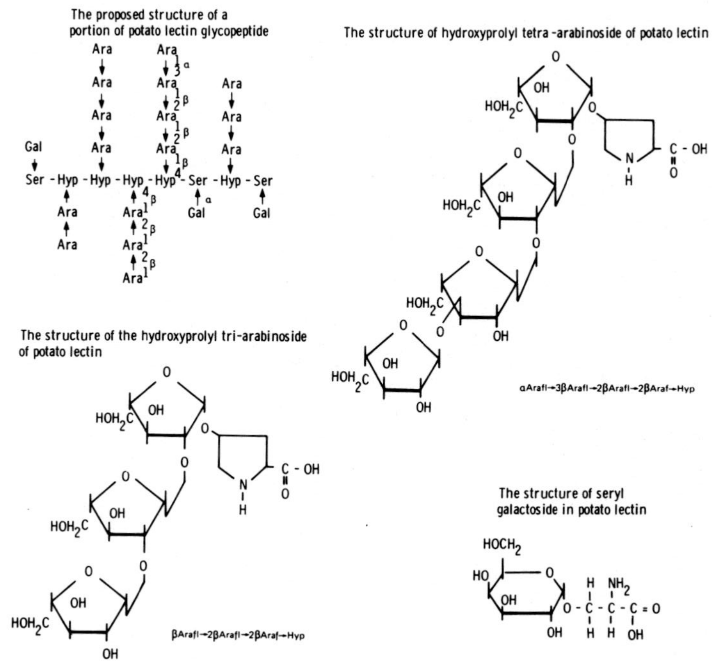

Fig. 1. Structure of the saccharides in the glycosylated domains of potato lectin.

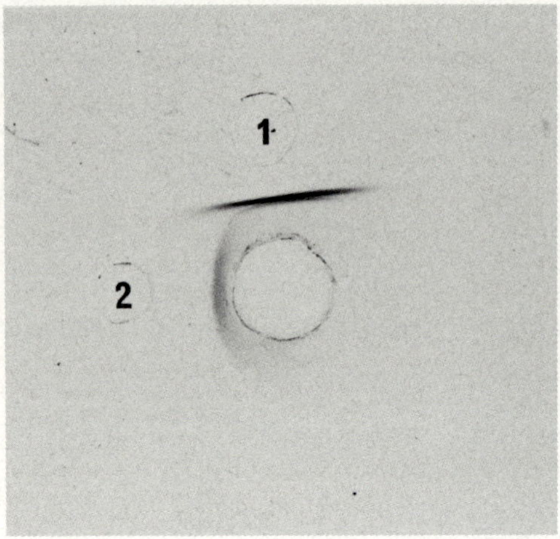

Fig. 2. Immunodiffusion of native potato lectin (1) and deglycosylated potato lectin (2) against rabbit anti-(potato lectin) serum (centre well).

THE BINDING SITES OF POTATO LECTIN

As has been shown above, the covalently bound carbohydrate and the associated hydroxyproline-rich domain is not involved in the saccharide binding site of potato lectin. It is therefore assumed that the binding sites for the N-acetylglucosamine oligosaccharides lie in the cystine-rich non-glycosylated region, and the disulphide bridges are essential in maintaining the structure of the binding sites since neither SCMC-lectin nor SAEC-lectin are active (Allen *et al.* 1978). Identification of the amino acids which are essential for binding of the saccharides was made (Ashford *et al.* 1981) by means of reagents that had a high degree of specificity for particular amino acid side chains. The reagents were also chosen because the reaction products could be determined quantitatively. Also, where possible, reagents were chosen where the modification produced could be reversed, regenerating the original amino acid residue. In cases where there was inactivation by a reagent the procedure was repeated in the presence of pseudo substrate tri-N-acetyl-chitotriose, and only if this saccharide protected the lectin from inactivation was it concluded that a particular residue was involved in the binding site. By these methods we concluded that lysine, arginine, histidine, methionine, cysteine, glutamic acid, and aspartic acid were not involved in the binding sites. Although we concluded that carboxyl and amino groups are not involved in the actual binding of substrate there is inactivation of the lectin after carbodiimide activation in the absence of pseudo substrate because of the formation of intramolecular cross links between carboxyl and amino groups. Tyrosine and tryptophan are undoubtedly involved in the binding sites as not only is the activity effected by specific chemical modifications, but u.v. difference spectra in the presence of chitin oligosaccharides have shown marked perturbations in appropriate regions of the spectrum (Ashford *et al.* 1982c). Equilibrium binding studies have suggested that there is only one multiple binding site per 50,000 subunit (Ashford *et al.* 1982d).

SPECIFICITY OF POTATO LECTIN

That the lectin is specifically inhibited by oligosaccharides which contain N-acetylglucosamine was first shown by Pardoe *et al.* (1969) and Masumoto and Osawa (1971). We later extended these observations (Table 3) (Allen, Neuberger 1973; Desai *et al.* 1981) to show that the monosaccharide is

Table 3. Comparison of the inhibitory effect of various saccharides on the agglutinating activity of potato lectin. Oligosaccharides contained β1,4-linkages.

Compound	Concentration needed for 50% inhibition (mM)
GlcNAc	200
Methyl α-GlcNAc	250
Methyl β-GlcNAc	60
$(GlcNAc)_2$	0.06
$(GlcNAc)_3$	0.03
$(GlcNAc)_4$	0.008
GlcNAc-MurNAc	8
$(GlcNAc\text{-}MurNAc)_2$	0.1
Fetuin glycopeptide	0.24
Soybean glycopeptide	0.24
$(Glc)_2$	>400

a weak inhibitor but there is some preference for the β-anomer, and for aromatic aglycones.

The inhibitory power of N-acetylated chito-oligosaccharides increases with increase in chain length up to the tetraose with a 20,000-fold increase in effectiveness. This suggests that there is an extended binding site with a number of subsites (Allen, Neuberger 1973) as in lysozyme and wheat-germ agglutinin (Allen *et al.* 1973). It is noteworthy that the bacterial cell wall oligosaccharides are weaker inhibitors than the chito-oligosaccharides. We have observed that glycopeptides are weaker inhibitors of the lectin than are the chito-oligosaccharides, an observation that was also made by Debray *et al.* (1981) for a number of other glycopeptides. This also agrees with our observations that complete glycoproteins are poor inhibitors of potato lectin and do not readily form precipitation lines in gel diffusions against potato lectin. We therefore conclude that the binding site of potato lectin binds most effectively to a tetrasaccharide of β1,4-linked N-acetylglucosamine residues.

AN ANTISERUM TO POTATO LECTIN AND ITS CROSS REACTIVITY WITH OTHER LECTINS FROM THE SOLANACEAE

We have prepared an antiserum to potato lectin (Ashford *et al.* 1982b). This antiserum was raised against lectin from 'King Edward' variety but gave a line of identity with purified lectins from, or extracts of, eight different varieties of potato. There are infact two lines of precipitation with the native lectin representing in our opinion two populations of antibodies, one against the glycosylated portion and one against the non-glycosylated portion of the lectin. As is shown in Fig. 2, deglycosylation of the lectin resulted in only one of the lines precipitating. In Fig. 3, it can be seen that the antiserum cross-reacted (with a line of identity) with the same lectin, the glycopeptide derived by pronase digestion, but not with the glycopeptide from which all the arabinose had been removed by acid treatment. The antiserum did not cross react with wheat-germ agglutinin, lysozyme, marrow lectin, polyhydroxyproline, arabinan, arabinogalactan or gum arabic. It did cross-react with two other solanaceous lectins (Fig. 4) from *Datura stramonium* (Jimson weed) and *Lycopersicon esculentum* (tomato). From partial precipitation studies with the glycopeptide we concluded that this cross reaction was with the glycosylated region, as both

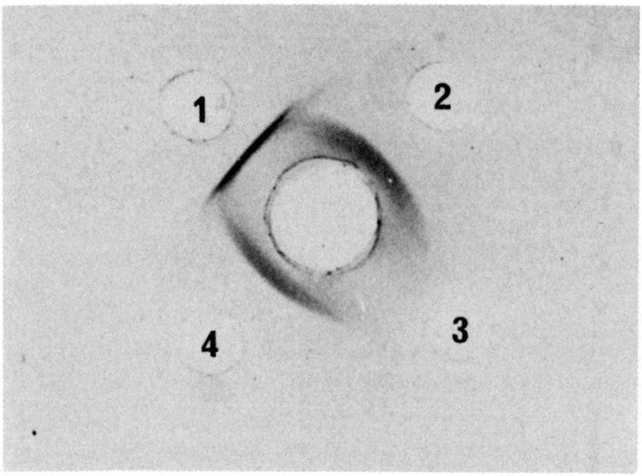

Fig. 3. Immunodiffusion of anti-potato lectin with potato lectin and its derivatives. Centre well, Anti(potato lectin) serum. (1) Native potato lectin. (2) Potato lectin glycopeptide. (3) Acid-treated glycopeptide. (4) SCMC-Potato lectin.

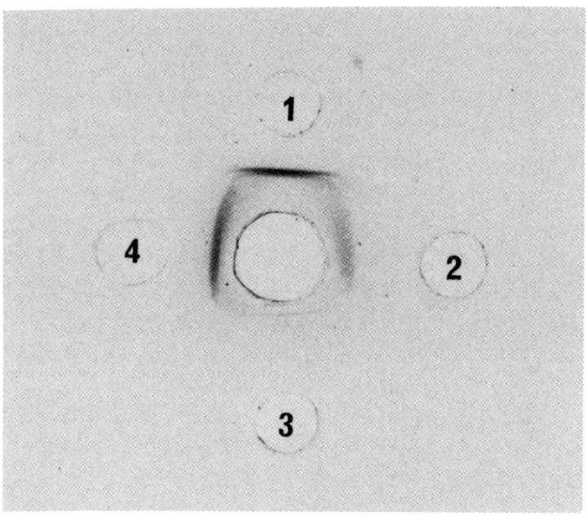

Fig. 4. Immunodiffusion of anti-potato lectin with related lectins. Centre well, Anti(potato lectin) serum. (1) Native potato lectin. (2) Tomato lectin. (3) *Datura stramonium* lectin. (4) *Datura* lectin glycopeptide.

lectins are known to contain hydroxyproline arabinosides. We suggested that the cross reactivities observed by Kilpatrick *et al.* (1980) between an anti(*Datura* lectin)serum and a number of solanaceous lectin extracts would have been against the carbohydrate.

CONCLUSION

From the investigations that have been done so far on potato lectin by various workers including those in our laboratory we can construct a hypothetical model for potato lectin (Fig. 5). It is likely to be an elongated molecule which exists in a monomer-dimer equilibrium. It will have two or more very dissimilar domains: one will be a binding-site domain with four subsites for saccharide binding which is rich in tyrosine and tryptophan residues and depends on disulphide bridges for its stability. The other region will have a hydroxyproline helix in which the hydroxyproline residues will almost all be substituted with tri- or tetra-arabinofuranosides interspersed with serine residues which are monosubstituted with galactopyranosides.

ACKNOWLEDGEMENTS

The work that I have reported here has been in association with my colleagues Dr. David Ashford, Dr. Nila Desai and Professor Albert Neuberger at Charing Cross Hospital Medical School, with Dr. Michael Creeth from Bristol, and Dr. Robert Selvendran and Mr. Malcolm O'Neill from Norwich. Various parts of the work have been funded by the Medical Research Council, the Agricultural Research Council and the Nuffield Foundation.

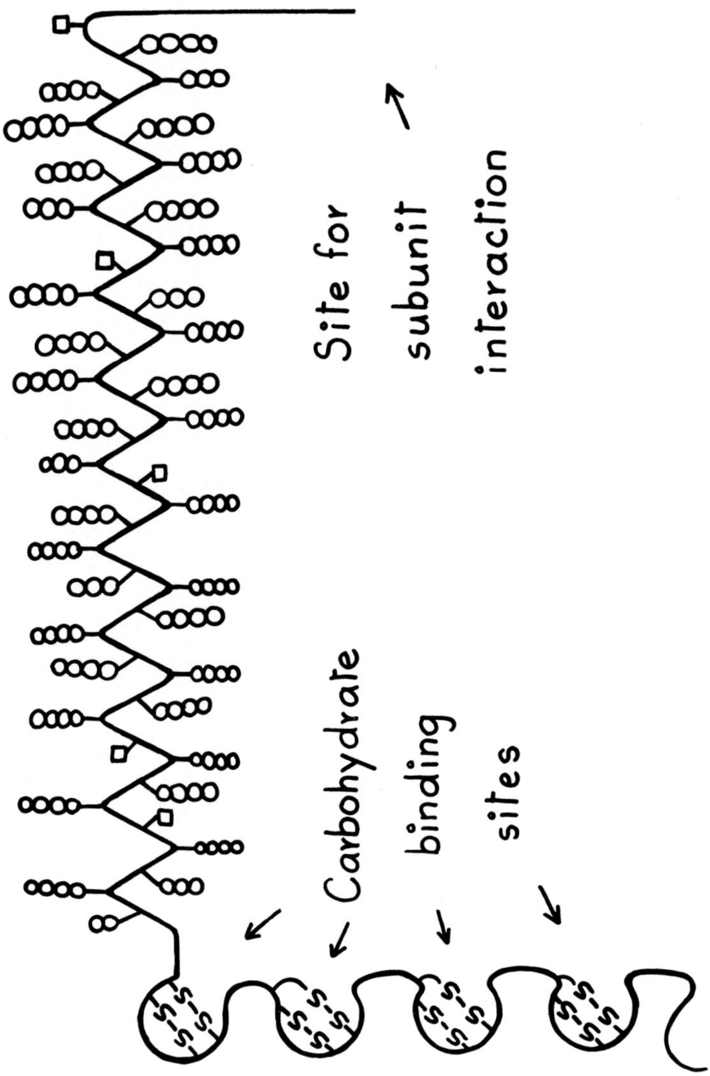

Fig. 5. Hypothetical model of the structure of potato lectin. O = Arabinofuranoside; □ = galactopyranoside.

REFERENCES

Akiyama Y, Mori M, Kato K (1980). ^{13}C-NMR analysis of hydroxyproline arabinosides from *Nicotiana tabacum*. Agric Biol Chem (Tokyo) 44: 2487.

Allen AK, Neuberger A (1973). The purification and properties of the lectin from potato tubers, a hydroxyproline-containing glycoprotein. Biochem J 135:307.

Allen AK, Neuberger A (1978). Potato lectin. Methods Enzymol 50C:340.

Allen AK, Neuberger A, Sharon N (1973). The purification, composition and specificity of wheat-germ agglutinin. Biochem J 131:162.

Allen AK, Desai NN, Neuberger A, Creeth JM (1978). Properties of potato lectin and the nature of its glycoprotein linkages. Biochem J 171:665.

Ashford D, Menon R, Allen AK, Neuberger A (1981). Studies on the chemical modification of potato (*Solanum tuberosum*) lectin and its effect on haemagglutinating activity. Biochem J 199:399.

Ashford D, Desai NN, Allen AK, Neuberger A, O'Neill MA, Selvendran RR (1982a). Structural studies of the carbohydrate moieties of lectins from potato (*Solanum tuberosum*) and thorn apple (*Datura stramonium*) seeds. Biochem J 201: 199.

Ashford D, Allen AK, Neuberger A (1982b). The production and properties of an antiserum to potato (*Solanum tuberosum*) lectin. Biochem J 201:641.

Ashford D, Allen AK, Neuberger A (1982c). Measurement of the interactions of potato (*Solanum tuberosum*) lectin with chito-oligosaccharides by ultra violet difference spectroscopy. Biochem Soc Trans 10:358.

Ashford D, Allen AK, Neuberger A (1982d). Measurement of the interactions of potato (*Solanum tuberosum*) lectin with chito-oligosaccharides by equilibrium dialysis. Proc 12th Int Cong Biochem (Perth, Australia) POS 004-248.

Debray H, Decout D, Strecker G, Spik G, Montreuil J (1981). Specificity of twelve lectins towards oligosaccharides and glycopeptides related to N-glycosylproteins. Eur J Biochem 117:41.

Delmotte F, Kieda C, Monsigny M (1975). Protein-sugar interaction by affinity chromatography of *Solanum tuberosum* (STA-lectin). FEBS Lett 53:324.

Desai NN, Allen AK (1979). The purification of potato lectin by affinity chromatography on an N,N',N"-triacetylchitotriose-Sepharose matrix. Anal Biochem 93:88.

Desai NN, Allen AK, Neuberger A (1981). Some properties of the lectin from *Datura stramonium* (thorn apple) and the nature of its glycoprotein linkages. Biochem J 197:345.

Desai NN, Allen AK, Neuberger A (1983). The properties of potato (*Solanum tuberosum*) lectin after deglycosylation by trifluoromethane-sulphonic acid. Biochem J 211:273.

Edge ASB, Faltnek CR, Hof L, Reichert LE, Weber P (1981). Deglycosylation of glycoproteins by trifluoromethane-sulfonic acid. Anal Biochem 118:131.

Gelhorn E (1925). in Schiff F (ed) "Handbuch der Biochemie des Menschen und Tiere", Gustav Fischer, Jena, Vol 2, p 346.

Jermyn MA, Yeow YM (1975). A class of lectins present in the tissues of seed plants. Aust J Plant Physiol 2:501.

Kilpatrick DC, Jeffree CE, Lockhart CM, Yeoman MM (1980). Immunological evidence for structural similarity among lectins from species of the Solanaceae. FEBS Lett 113:129.

Krüpe M (1956). "Blutgruppenspezifische Pflanzliche Eiweisskörper", Enke, Stuttgart, p 11.

Marcusson-Begun H (1925). Untersuchungen über das Hämagglutinin der Kartoffelknolle. Z Immunitaetsforsch Exp Ther 45:49.

Marinkovich VA (1964). Purification and characterization of the hemagglutinin present in potatoes. J Immunol 93:732.

Masumoto I, Osawa T (1971). On the specificity of various heterologous anti-H hemagglutinins. Vox Sang 21:548.

Masumoto I, Mizuno Y, Seno N (1979). Activation of Sepharose with epichlarohydrin and subsequent immobilization of ligand for affinity adsorbent. J Biochem (Tokyo) 85:1091.

Masumoto I, Jimbo A, Mizuno Y, Seno N, Jeanloz RW (1983). Purification and characterization of potato lectin. J Biol Chem 258:2886.

Muray RHA, Northcote DH (1978). Oligoarabinosides of hydroxyproline isolated from potato lectin. Phytochemistry 17:623.

Owens RJ, Northcote DH (1981). The location of arabinosyl: hydroxyproline transferase in the membrane system of potato tissue culture cells. Biochem J 195:661.

Pardoe GI, Bird GWG, Uhlenbruck G (1969). On the specificity of lectins with a broad agglutination spectrum. I. The nature of the specific receptors for *Ricinus communis* and *Solanum tuberosum* lectins. Z Immunitaetsforsch Allerg Klin Immunol 137:442.

Singh G, Verma SD, Bird GWG (1961). Studies on the haemagglutinins present in potatoes (*Solanum tuberosum*). Z Immunitaetsforsch Exp Ther 121:181.

Chemical Taxonomy, Molecular Biology, and Function of Plant Lectins, pages 87–98

STRUCTURE AND EXPRESSION OF SOYBEAN LECTIN GENES

Lila O. Vodkin

Plant Genetics and Germplasm Institute
U. S. Department of Agriculture
Beltsville, Maryland 20705

INTRODUCTION

A summary of recent work on the molecular analysis of a naturally occurring mutation which affects the expression of soybean lectin (SBL) is presented. The analysis of induced or natural mutants in eukaryotes has lead to substantial insights into gene structure and regulation. Primarily, these studies have been conducted in animal systems; however, the ease and widespread use of recombinant DNA techniques is presently encouraging more interest in plant gene expression and development.

At the beginning of these studies, there were several lines of evidence indicating that SBL would be a good system to use in molecular approaches toward understanding gene structure and function in plants. SBL is a moderately abundant protein, generally accounting for 2-5% of the total seed protein at maturity. The biochemical properties of soybean lectin have been extensively characterized. It is a 120,000 dalton tetramer of apparently identical subunits, each containing about 2000 daltons of attached carbohydrate (Lotan, et.al., 1974). Thus, SBL is a simpler system than glycinin and conglycinin, the major storage proteins of the soybean seed. Glycinin, for example, is a 360,000 dalton complex consisting of different acidic and basic polypeptide subunits (Barton, et.al., 1982).

More importantly, a screen of soybean cultivars and plant introductions in the USDA soybean germplasm collection revealed several lines that lacked detectable lectin protein (Pull, et.al., 1978). The segregation patterns of crosses between cultivars which express lectin (Le^+ phenotype) and those which do not (Le^- phenotype) indicated that absence of seed lectin follows a simple recessive mode of inheritance (Orf, et.al., 1979). Thus, our objective in the present studies has been to determine the molecular reason for the inherited absence of lectin protein in certain soybean cultivars. One can envision several possibilities including translational defects, nuclear RNA processing defects, or a total or partial deletion of the lectin gene. Surprisingly, we have found that a large insertion of DNA which has properties resembling a mobile element is the cause of the lectinless phenotype.

EXPRESSION OF SEED LECTIN TRANSCRIPTS IN LE^+ AND LE^- PLANTS

Lectin protein accumulates in the cotyledons during the mid-maturation phase of soybean embryo development. Lectin mRNA was preparatively isolated from Le^+ immature cotyledons of approximately 100-300 mg fresh weight per seed by a polysome immunoadsorption procedure. The details have been presented elsewhere (Vodkin, 1981). Briefly, polysomes were isolated from the embryos and incubated with monospecific antibody produced against SDS-denatured lectin protein. The antibody tags nascent lectin polypeptides on the polyribosomes which are then selectively removed by chromatography on a Sepharose affinity column containing a second antibody (goat anti-rabbit antibody, for example) or protein A. After washing, the bound material was eluted with EDTA which dissociates the polyribosome complex. Poly A^+ RNA was isolated from the immunoadsorbed polysomes by two cycles of passage over oligo-dT cellulose.

In vitro translation was used to analyse the degree of enrichment of lectin mRNA. Figure 1 demonstrates that a major polypeptide of 32,300 daltons and a minor species of 29,000 are synthesized by lectin mRNA in a rabbit reticulocyte system. Both of these polypeptides are immunoprecipitable by monospecific lectin antibody (data not shown). Compared to the *in vitro* products produced by total mid-maturation mRNA (Figure 1A, lane 4), it is evident that there has been a substantial enrichment of the lectin message by the polysome selection procedure. The

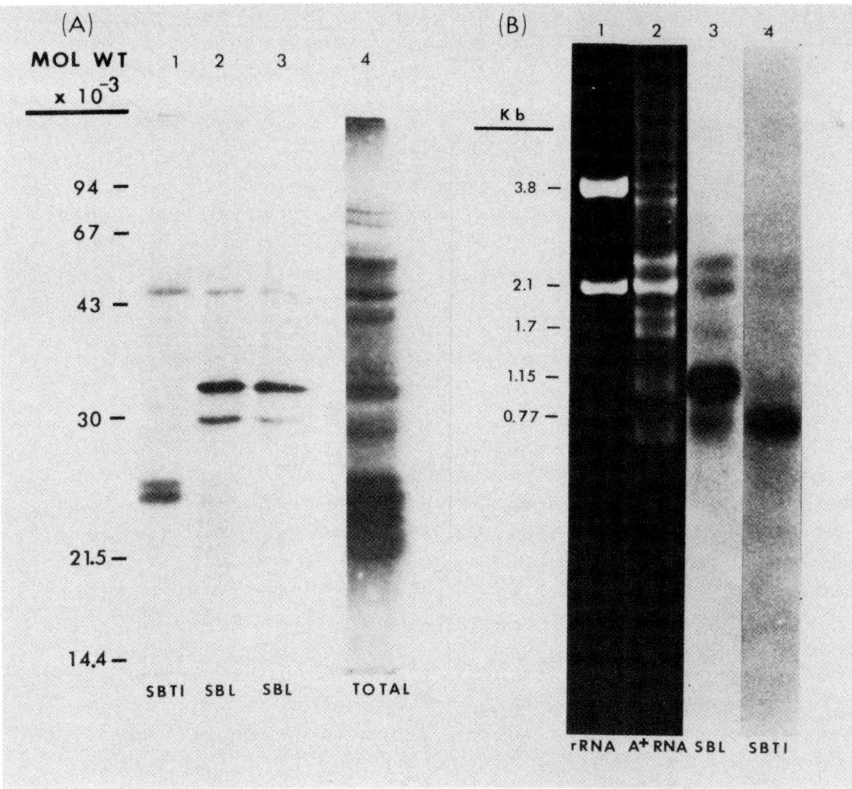

Fig. 1A. Analysis of translation products by SDS/polyacrylamide gel electrophoresis and fluorography (from Vodkin, 1981). Lane 1, 60 ng of enriched soybean trypsin inhibitor mRNA; lanes 2 and 3, 60 ng of enriched SBL mRNA; lane 4, 500 ng of total mRNA. The band at approximately 45,000 daltons is a contaminant polypeptide synthesized from endogenous RNA in the rabbit reticulocyte system. Fig. 1B. Northern blot analysis of lectin and trypsin inhibitor mRNAs. Total mRNA from soybean embryos was electrophoresed on a 1% agarose gel in the presence of 5 mM methymercury hydroxide denaturant. The mRNAs were transferred to DPT-paper strips and hybridized to ^{32}P-labeled cDNA prepared using immunoselected SBL or trypsin inhibitor mRNAs. Lanes 1 and 2, rRNA and mRNA profiles as visualized by ethidium bromide staining; lanes 3 and 4, hybridization of labeled SBL and trypsin inhibitor cDNAs to the total mRNA profile as shown in lane 2.

in vitro translation data indicate that SBL is synthesized with a 3000-4000 dalton precursor signal sequence. The 29,000 dalton polypeptide is likely a premature termination product or a proteolytic modification of the major polypeptide.

In another test to determine the degree of purity, the lectin message was reverse transcribed and the labeled cDNA was used to probe blots of total mRNA from soybean embryos. As shown in Fig. 1B, the lectin cDNA hybridized predominantly to a message of approximately 1.1 kb in size. The immunoselection procedure was used to enrich the Kunitz soybean trypsin inhibitor mRNA also, as illustrated in Figures 1A and 1B.

To obtain a cDNA clone, we utilized the enriched lectin message as a probe to screen a cDNA bank. The bank was constructed in pBR322 using cDNAs prepared from total soybean mid-maturation messages (Goldberg, *et.al.*, 1981). A putative lectin clone, designated L9, was confirmed to be lectin by hybrid-select translation (Hoschek and Goldberg, unpublished) and by direct sequence analysis (Vodkin and Rhodes, unpublished).

Having a cloned probe, we were able to investigate the expression of the seed lectin transcripts in Le^+ and Le^- embryos in detail. Figure 2 shows that the pattern of ethidium bromide stained mid-maturation mRNAs was similar in both Le^+ and Le^- embryos. When the mRNAs were transferred to paper and probed with the labeled L9 lectin plasmid, we found that Le^- embryos did not have detectable lectin sequences. As a control on the RNA preparation from the Le^- cotyledons, a different seed protein cDNA clone was included in the hybridization reaction. It hybridized with approximately equal intensity to both Le^+ and Le^- mRNA populations.

More detailed studies on the quantitative levels of lectin transcripts showed that lectin messages are found at approximately 4000 molecules per cell in the cytoplasm of normal embryos and are reduced 10,000 fold in Le^- embryos (Goldberg, *et.al.*, 1983). Similarly, lectin transcripts in the nuclear RNA fraction are reduced drastically. These results suggest that the Le^- phenotype is the result of a transcriptional defect rather than disruption of translational or RNA processing events.

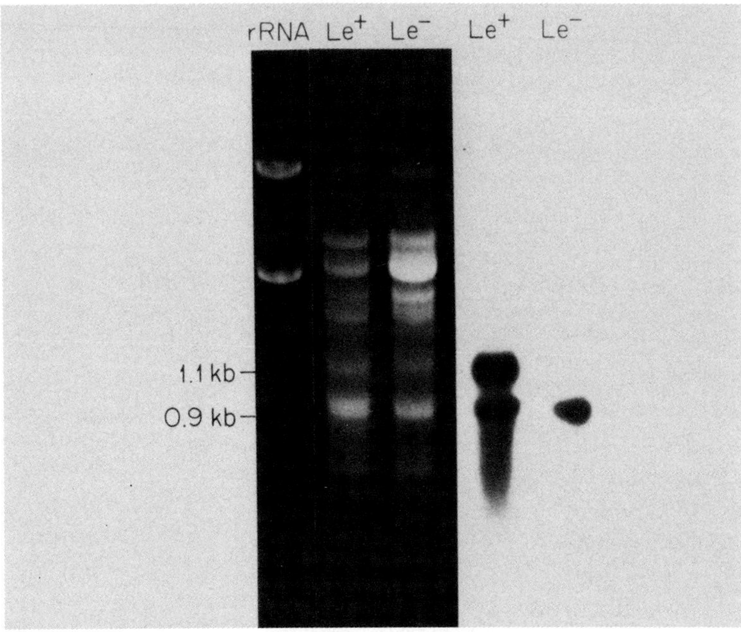

Fig. 2. Hybridization of Le^+ and Le^- embryo RNAs with lectin plasmid L9 (adapted from Goldberg, et.al., 1983). Mid-maturation mRNAs from Le^+ and Le^- embryos were separated by electrophoresis, blotted to paper, and hybridized with labeled L9 and A36 plasmid probes. A36 is a clone of a prevalent 15 kd seed protein and is included as a positive control on the Le^- RNA preparation. It hybridizes to a 0.9 kb message in both Le^+ and Le^- mRNA preparations.

STRUCTURE OF LECTIN GENES

To determine whether the seed lectin gene is present in the lectinless cultivar, DNAs were extracted from the leaves of both Le^+ and Le^- plants, digested with Eco RI and compared by the Southern blotting procedure using the L9 lectin probe. As shown in Figure 3A, the patterns were identical in both the Le^+ and Le^- plants. L9 hybridized most intensely to a 17 kb Eco RI restriction fragment and less intensely to a 5.4 kb Eco RI fragment in each of the DNA types.

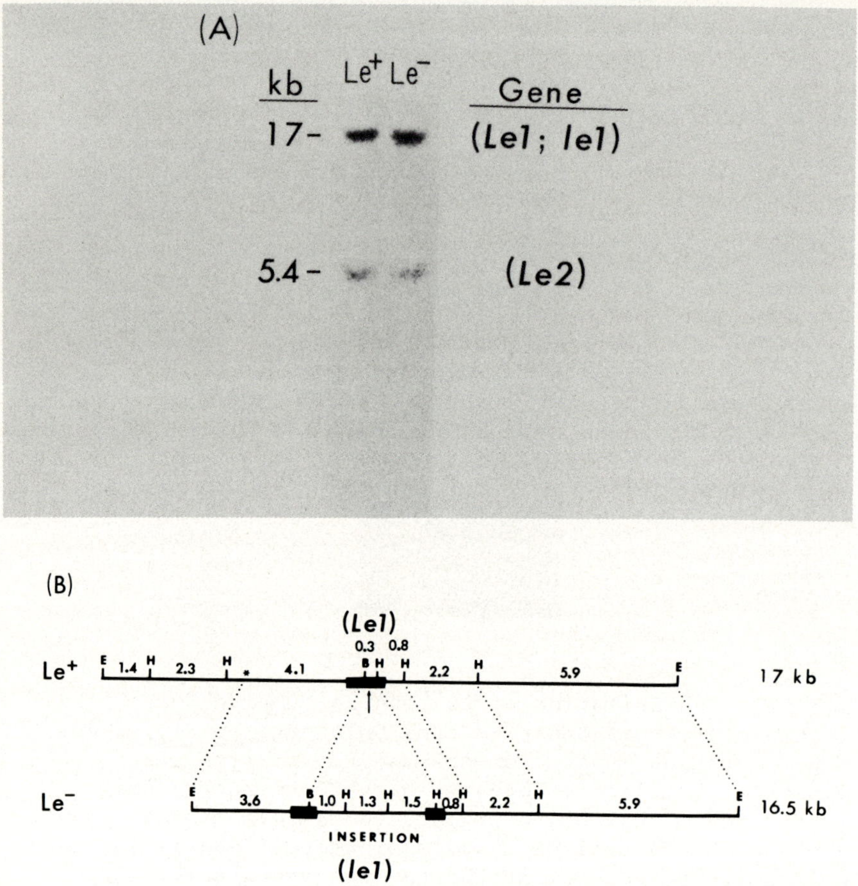

Fig. 3A. Comparison of lectin genes in Le^+ and Le^- plants (adapted from Goldberg, et.al., 1983). Soybean genomic DNAs were digested with Eco RI, electrophoresed on 0.8% agarose, transferred to nitrocellulose, and hybridized with labeled L9 plasmid DNA.

Fig. 3B. Comparison of the Eco RI (E), Hind III (H), and Bam HI (B) restriction sites on the Le^+ and Le^- cloned gene regions. The lectin coding regions are indicated by the thick horizontal line. Corresponding genomic regions are connected by the dotted lines. The Eco RI site polymorphism is indicated by an asterisk, and the point of insertion is denoted by the vertical arrow.

The genomic clones which correspond to all four of these restriction fragments have been isolated from bacteriophage lambda libraries and characterized (Goldberg, et.al., 1983). In summary, the 17 kb fragment from Le^+ DNA contains the normal, expressed seed lectin gene (designated Le1) and the 17 kb Eco RI fragment from Le^- DNA contains the non-expressed recessive allele (le1). The 5.4 kb Eco RI fragments from both Le^+ and Le^- DNAs contain a divergent lectin gene (designated Le2) which will be discussed later.

The near identity of the 17 kb seed lectin genes, as seemed apparent from the Southern blots of Fig. 3A, was a deceptive result. Heteroduplex studies and detailed restriction mapping of the clones showed that these genomic regions differed in two major respects. These differences are summarized schematically in Figure 3B. The surprising and important result was the finding that a large insertion of extra DNA (3.4 kb) interrupted the lectin coding region in the le1 allele. The second difference was an Eco RI restriction site polymorphism that resulted in the Le^- clone lacking a 4.2 kb Eco RI - Hind III restriction fragment at its 5' end. Therefore, despite the 3.4 kb insertion the le1 allele, the Le^- Eco RI fragment is approximately the same size as the Le^+ fragment.

We have compared the normal and mutant genes in more detail. As shown in Figure 4, several restriction sites in these genes are identical with the exception that one of the lectin Ava I sites is not found in le1. For clarity in Figure 4, the 3.4 kb insertion DNA has been omitted but its position nearly two thirds into the lectin gene is indicated. The borders of the insertion element were sequenced and found to be near-perfect inverted repeats of 35 base pairs (data not shown). In addition, a three base sequence of lectin DNA which occurs only once in Le1 has been duplicated in le1 so that it flanks each end of the insertion. Thus, by these two characteristics, the lectin insertion element resembles mobile elements in other systems. The presence of inverted repeat termini and of small flanking duplications of target DNA are characteristic features of transposable elements.

The complete nucleotide sequence of Le1 demonstrates that it encodes a 1.0 kb mRNA and lacks introns (data not shown). The corresponding lectin regions in the mutant gene were also sequenced and compared to Le1. In all, there are six single base substitutions between the two alleles. Two occur in about 200 bp of flanking 5' sequence, three in the coding region of the gene, and one in 200 bp of 3' flanking sequence. Both genes have the same consensus promoter sequence found 27 bases upstream of the message 5' end. In summary, the major difference between the expressed seed lectin gene and the non-expressed allelic form is the 3.4 kb insertion element. We conclude that the insertion prevents transcription from the 5' end of the le1 allele and leads to the lectinless phenotype in Le$^-$ plants.

Fig.4. Comparison of restriction sites in the lectin genes and flanking regions. Le1 and le1 are allelic forms of the seed lectin gene. The dotted line indicates the message coding region of the gene. The 3.4 kb insertion in the le1 allele has been omitted for clarity but its position is indicated. Le2 is a divergent relative of the seed lectin gene. Asterisks denote restriction site polymorphisms. The partial cDNA clone, L9 (Le1 gene), is also indicated.

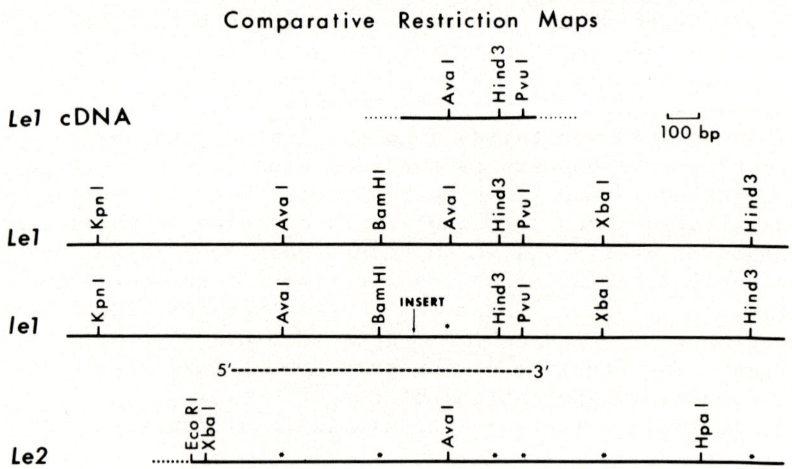

The combined evidence from molecular, genetic, and biochemical data demonstrates that the production of soybean seed lectin is under the control of one structural gene locus, Le1. Other seed lectins may not have as simple a genetic basis as the soybean lectin since biochemical studies indicate the presence of multiple lectins or dissimilar lectin subunits in a number of plants. Although SBL is encoded by a single gene locus, we have identified a second gene, designated Le2, which has some homology to the seed lectin gene as indicated in Figure 3. Many of its restriction sites are different from the Le1 locus (see Fig. 4) and initial nucleotide sequence analysis of certain regions indicates about 85% homology. The Le2 gene is present in both Le^+ and Le^- cultivars. Since lectin transcripts are greatly reduced in Le^- cotyledons, it is clear that an Le2 product does not contribute to the production of seed lectin. We are currently assaying for the presence of an mRNA and protein product from Le2 in other plant parts. Preliminary results indicate that it is expressed at low levels in the roots.

SOYBEAN LECTIN PROTEIN STRUCTURE

SBL has been partially sequenced at the protein level (Hemperly, et.al., 1982). The complete amino acid sequence of soybean lectin as deduced from the DNA data is shown in Figure 5. The sequence for the mature lectin protein is preceded by a hydrophobic signal peptide of 32 amino acids (3,360 daltons) as was expected from the *in vitro* translation data presented in Figure 1. SBL is known to be localized, at least in part, in the protein bodies of the cotyledons (Horisberger and Vonlanthen, 1980). A subunit of the *Phaseolus vulgaris* lectin, PHA-E, contains a 22-amino acid signal sequence, as deduced from the sequence of a cDNA plasmid (Hoffman, et.al., 1982). Likewise, favin, the lectin from *Vicia faba*, is synthesized with a 29-amino acid signal (Hemperly, et.al., 1982). Signal sequences have been demonstrated at the nucleotide level for zein and phaseolin, the major storage polypeptides in maize and the French bean, respectively (Pedersen, et.al., 1982; Slightom, et.al., 1983). Thus, it appears that processing of seed lectins which are found in protein bodies may be similar to that of the major storage proteins.

Fig. 5. The complete amino acid sequence of soybean lectin. Data are derived from the nucleotide sequence of Le1 from the cultivar Forrest (adapted from Vodkin, *et.al.*, in preparation). The 32 residue signal sequence is indicated by lower case letters.

									signal	peptide	=		met	ala
thr	ser	lys	leu	lys	thr	gln	asn	val	val	val	ser	leu	ser	leu
thr	leu	thr	leu	val	leu	val	leu	leu	thr	ser	lys	ala	asn	ser
1									10					
ALA	GLU	THR	VAL	SER	PHE	SER	TRP	ASN	LYS	PHE	VAL	PRO	LYS	GLN
				20										30
PRO	ASN	MET	ILE	LEU	GLN	GLY	ASP	ALA	ILE	VAL	THR	SER	SER	GLY
									40					
LYS	LEU	GLN	LEU	ASN	LYS	VAL	ASP	GLU	ASN	GLY	THR	PRO	LYS	PRO
				50										60
SER	SER	LEU	GLY	ARG	ALA	LEU	TYR	SER	THR	PRO	ILE	HIS	ILE	TRP
									70					
ASP	LYS	GLU	THR	GLY	SER	VAL	ALA	SER	PHE	ALA	ALA	SER	PHE	ASN
				80										90
PHE	THR	PHE	TYR	ALA	PRO	ASP	THR	LYS	ARG	LEU	ALA	ASP	GLY	LEU
									100					
ALA	PHE	PHE	LEU	ALA	PRO	ILE	ASP	THR	LYS	PRO	GLN	THR	HIS	ALA
				110										120
GLY	TYR	LEU	GLY	LEU	PHE	ASN	GLU	ASN	GLU	SER	GLY	ASP	GLN	VAL
									130					
VAL	ALA	VAL	GLU	PHE	ASP	THR	PHE	ARG	ASN	SER	TRP	ASP	PRO	PRO
				140										150
ASN	PRO	HIS	ILE	GLY	ILE	ASN	VAL	ASN	SER	ILE	ARG	SER	ILE	LYS
									160					
THR	THR	SER	TRP	ASP	LEU	ALA	ASN	ASN	LYS	VAL	ALA	LYS	VAL	LEU
				170										180
ILE	THR	TYR	ASP	ALA	SER	THR	SER	LEU	LEU	VAL	ALA	SER	LEU	VAL
									190					
TYR	PRO	SER	GLN	ARG	THR	SER	ASN	ILE	LEU	SER	ASP	VAL	VAL	ASP
				200										210
LEU	LYS	THR	SER	LEU	PRO	GLU	TRP	VAL	ARG	ILE	GLY	PHE	SER	ALA
									220					
ALA	THR	GLY	LEU	ASP	ILE	PRO	GLY	GLU	SER	HIS	ASP	VAL	LEU	SER
				230										240
TRP	SER	PHE	ALA	SER	ASN	LEU	PRO	HIS	ALA	SER	SER	ASN	ILE	ASP
									250					
PRO	LEU	ASP	LEU	THR	SER	PHE	VAL	LEU	HIS	GLU	ALA	ILE	stop	

Soybean lectin shows a large degree of homology to other seed lectins (see Strosberg, this volume). An interesting permutation in the protein sequence between concanavalin A and other legume lectins has been noted (Cunningham, et.al., 1979). Soybean lectin is also circularly permuted with respect to con A as are favin and lentil lectin (see Hemperly, this volume). The N-terminal sequence of con A corresponds to the sequence beginning at amino acid 116 of soybean lectin; likewise the C-terminal of con A ends near amino acid 115 of soybean lectin. The origin of this permuted structure is unknown. Intron rearrangements in an ancestral species is unlikely to be the cause since the soybean lectin gene does not contain introns.

CONCLUSIONS

Molecular data showing that soybean seed lectin is produced by a single gene (Le1) have been summarized. We have also determined that certain cultivars which do not express lectin protein possess an allelic form (le1) of the seed lectin gene that contains a 3.4 kb insertion element. This element interrupts the coding region of the gene and prevents transcription of 5' lectin sequences. It has the structural features of a transposable element. Clearly, there are many questions to be answered concerning the structure and distribution of this element in the genome and its effect on gene expression.

These studies have focused primarily on molecular approaches to gene structure and expression using soybean seed lectin as an initial system for study. However, the amino acid sequence of soybean lectin as deduced from the nucleotide sequence, has contributed to the evolutionary taxonomy and structural analyses of lectins. Soybean seed lectin apparently has no critical function in the plant since its total absence in Le^- cultivars creates no observable phenotypic differences in these plants. Whether a physiological role for a product of the Le2 gene in root tissue can be defined awaits further study.

Acknowledgements: I thank my associate, Patsy Rhodes, for reviewing the manuscript. The results summarized here represent a collaborative effort with the laboratory of Robert Goldberg, Biology Dept., Univ. Calif., Los Angeles.

REFERENCES

Barton KA, Thompson JF, Madison JT, Rosenthal R, Jarvis NP, Beachy, RN (1982). The biosynthesis and processing of high molecular weight precursors of soybean glycinin subunits. J Biol Chem 257: 6089.

Cunningham BA, Hemperly JJ, Hopp TP, Edelman GM (1979). Favin versus concanavalin A: Circularly permuted amino acid sequences. Proc Natl Acad Sci USA 76: 3218.

Goldberg RB, Hoschek G, Ditta GS, Briedenbach RW (1981). Developmental regulation of cloned superabundant embryo mRNAs in soybean embryogenesis. Devel. Biol. 83: 201.

Goldberg RB, Hoschek G, Vodkin LO (1983). An insertion sequence blocks the expression of a soybean lectin gene. Cell 33: 465.

Hemperly JJ, Mostov KE, Cunningham, BA (1982). In vitro translation and processing of a precursor form of favin, a lectin from Vicia faba. J Biol Chem 257: 7903.

Hemperly JJ, Becker JW, Cunningham BA (1982). Plant lectins. In Bradshaw, et.al. (ed): "Proteins in Biology and Medicine" New York: Academic Press, p395.

Hoffman LM, Ma Y, Barker, R (1982). Molecular cloning of Phaseolus vulgaris lectin mRNA and use of cDNA as a probe to estimate lectin transcript levels in various tissues. Nuc Acid Res 10: 7819.

Horisberger M, Vonlanthen M (1980). Ultrastructural localization of soybean agglutinin on thin sections of Glycine max (soybean) var. Altona by the gold method. Histochem 65: 181.

Lotan R, Siegelman HW, Lis H, Sharon N (1974). Subunit structure of subunit agglutinin. J Biol Chem 249: 1219.

Orf JH, Hymowitz T, Pull SP, Pueppke SG (1979). Inheritance of a soybean seed lectin. Crop Sci 18: 899.

Pedersen K, Devereux J, Wilson DR, Sheldon E, Larkins BA, (1982). Cloning and sequence analysis reveal structural variation among related zein genes in maize. Cell 29: 1015.

Pull SP, Pueppke SG, Hymowitz T, Orf JH (1978). Soybean lines lacking the 120,000 dalton seed lectin. Science 200: 1277.

Slightom, JL, Sun SM, and Hall TC (1983). Complete nucleotide sequence of a French bean storage protein gene: Phaseolin. Proc Natl Acad Sci USA 80: 1897.

Vodkin LO (1981). Isolation and characterization of messenger RNAs for seed lectin and Kunitz trypsin inhibitor in soybeans. Pl Physiol 68: 766.

Chemical Taxonomy, Molecular Biology, and Function of Plant Lectins, pages 99–116

GRAMINEAE LECTINS: OCCURRENCE, MOLECULAR BIOLOGY AND PHYSIOLOGICAL FUNCTION.

Willy J. Peumans and Hetty M. Stinissen

Laboratorium voor Plantenbiochemie, KULeuven
Kardinaal Mercierlaan, 92
B-3030 Leuven, Belgium

INTRODUCTION

Ever since their discovery almost a century ago by Stillmark, plant lectins fascinated scientists in diverse disciplines and contributed markedly to important progresses in medicine, immunology, cancerology, biochemistry and cell-biology. Surprisingly, however, these typical plant proteins played a very inconspicuous role in botanical sciences and it is only during the last decades that phytohemagglutinins have been 'discovered' by plant scientists as interesting proteins and promising model systems.

As a result of the discrepancy between the rapid progress in the biochemistry and fysico-chemistry of plant lectins and the poor understanding of their biology (in a broad sense), the field of phytohemagglutinins has been dominated by the somewhat contradictory situation that no physiological role could be attributed with certainty to a group of well characterized proteins. In the past, several types of functions have been speculated for plant lectins, but at present, little conclusive evidence has been presented for any of them. It can be expected, however, that in the near future in-depth studies of botanical, molecular-biological and genetic aspects of different classes of phytohemagglutinins, including these that occur exclusively in vegetative plant tissues, will lead to a dramatic progress in the understanding of what plant lectins do and how they act.

The aim of this contribution is to give the state of the art for a well-defined class of phytohemagglutinins, namely the *Gramineae* lectins. Several lines of evidence, based on the results of biochemical, botanical, chemotaxonomic, genetical and molecular-biological studies, indicate a specific endogenous role for these lectins. The results of recent developments in the molecular-biology of some of these lectins further suggest that they might constitute a system, or at least part of a system, that controls the development of cereal embryos during seed formation and embryogenesis.

OCCURRENCE, STRUCTURAL SIMILARITIES AND IMMUNOLOGICAL RELATIONSHIPS OF CEREAL LECTINS

Wheat Germ Agglutinin: the Prototype of Cereal Lectins

Wheat germ agglutinin (WGA) is one of the first lectins that has been purified (Nagata, Burger 1972) and characterized in detail for what concerns its biochemical and physicochemical properties (Allen *et al*. 1973; Nagata, Burger 1974; Rice, Etzler 1974; Wright 1980). It behaves like a dimeric protein of molecular weight 36,000 composed of two non-covalently bound identical subunits (molecular weight 18,000). It is devoid of covalently bound carbohydrate and has a very high content of half-cystine (but no free sulfhydryl groups) and glycine. WGA is not blood-group-specific and exhibits carbohydrate-binding specificity for N-acetylglucosamine and oligomers of this sugar.

Although commonly used, the term wheat germ agglutinin is somewhat misleading since this lectin does not represent a single molecular species but consists of a mixture of isolectins which are structurally and functionally closely related (Peumans *et al*. 1982c). A genetic basis for the occurrence of multiple forms of WGA has already been proposed in 1976 (Rice 1976), but it is only recently that a definitive explanation has been presented (Peumans *et al*. 1982c). Analyses of isolectin patterns in *Triticum monococcum* (genome AA), *Triticum tauschii* (genome DD), *Triticum turgidum diccocum* (genome AABB), *Triticum timopheevii timopheevii* (genome AAGG), and *Triticum aestivum* (genome AABBDD) indicated that each genome, either in simple or complex (allopolyploid) genomes, directs the synthesis of a single lectin subunit species. In diploids, a single subunit species is present so that dimeric lectin molecules are exclusively formed by the association of identical subunits giving rise to only one molecular form. In polyploids, however, the different lectin polypeptide chains, which are coded for by the different genomes, combine fully randomly with identical and non-identical partners, whereby both homomeric and heteromeric dimers are formed in all possible combinations. As a result, tetraploid and hexaploid wheats contain 3 and 6 isolectins respectively. In order to avoid ambiguities in the designation of the different isolectins we propose to indicate them by the abbreviation CL (from cereal lectin), followed by a subscript indicating the genome(s) by which the subunits are coded for. Applied to haxaploid wheat this gives us CL_{AA}, CL_{BB}, CL_{DD}, CL_{AB}, CL_{AD} and CL_{BD}. The genetic origin of the different lectin subunits could be determined unequivocally by comparing the (iso)lectins of *Triticum monococcum*, *Triticum turgidum dicoccum*, *Triticum tauschii* and *Triticum aestivum*. In addition to an explanation for the complex and unique isolectin composition of WGA, these results also indicate that the lectin polypeptide chains coded for by the (related) A, B and D genomes are so closely related structurally that they cannot distinguish between identical and non-identical partners during their assembly into dimers.

WGA-Like Lectins from Rye and Barley

WGA has for a long time been considered as a unique lectin but evidence is now accumulating that it just represents a (complex) example of an extensive class of closely related proteins. Two lectins were isolated from rye and barley embryos respectively, with chemical, physical and biological properties almost identical to those of WGA (Peumans *et al.* 1982b). They do not only have the same molecular structure, specificity, dissociation characteristics and amino acid composition as WGA, but are also immunologically indistinguishable from WGA when challenged with WGA-antiserum. These striking structural similarities between the lectins of wheat, rye and barley are further emphasized by their ability to exchange subunits whereby intergeneric heteromeric lectins are formed (Peumans *et al.* 1982d).

Occurrence of WGA-Like Lectins within the Triticeae Tribe

In order to find out whether the occurrence of WGA-like lectins is confined to wheat, rye and barley, or whether they occur generally within the Triticeae tribe, we checked a number of species of this taxonomic group for the presence of WGA-like lectins(s). We found that embryos of all available Triticeae species (Table 1) contained without any exception N-acetylglucosamine-specific lectins that were immunologically indistinguishable from WGA (their precipitin lines fused completely with that of WGA upon double immunodiffusion against WGA-antiserum) (Stinissen *et al.* 1983a). Therefore, it appears that within the Triticeae tribe, which comprises

Table 1. List of Triticeae species containing embryo lectins immunologically indistinguishable from WGA.

Subtribe Triticinae		
Aegilops geniculata	*Agropyrum sibiricum*	*Triticum speltoides*
ligustica	*trachycaulon*	*tauschii*
markgrafii	*trichophorum*	*tripsacoides*
neglecta	*uganicum*	*umbellulatum*
peregrina	*Eremopyrum buonopardis*	*uniaristatum*
Agropyrum aucheri	*Haynaldia villosa*	*columnaris*
caninum	*Heteranthelium piliferum*	*crassa*
cristatum	*Secale cereale*	*cylindricum*
desertorum	*montanum*	*kotschyi*
elongatiforme	*kuprijanovii*	*macrochaetum*
elongatum	*silvestre*	*ovatum*
intermedium	*vavilovii*	*syriacium*
junceum	*Triticosecale*	*timopheevii*
litorale	*Triticum bicorne*	*turgidum*
panormitanum	*comosum*	*triuncale*
pauciflorum	*longissimum*	*variabilis*
pectineforme	*monococcum*	*aestivum*
repens	*searsii*	*juvenalis*
scabriglume	*sharonensis*	
Subtribe Hordeinae		
Elymus canadensis	*Hordeum jubatum*	*Hordeum violaceum*
doluricus	*marinum*	*vulgare*
glaucus	*maritimum*	*chilense*
sibiricus	*murimum*	*compressum*
Hordeum bogdanii	*nodosum*	*secalinum*
brevisbulatum	*spontaneum*	*leponicum*
bulbosum	*turkestanicum*	*Psathyrostachys juncea*

a considerable number of distinct genera, all lectins are immunologically indistinguishable and thus represent a class of very closely related proteins, which can be designated by the common name 'cereal lectins'.

WGA-Related Lectins in *Brachypodium* and *Oryza* Species

In order to find out whether *Gramineae* species, other than the Triticeae, also contain WGA-like or WGA-related lectins, we made a (partial) survey of the occurrence, sugar-binding specificity and possible cross-reactivity against WGA-antiserum of lectins within the grass family. Thereby, it became evident that besides the Triticeae only species belonging to the genera *Brachypodium* and *Oryza* (rice) contain without any exception lectins in their embryos (Table 2). In addition, all these lectins not only displayed the same carbohydrate-specificity as WGA, but also cross-reacted with WGA-antiserum (Stinissen *et al.* 1983a).

A more detailed analysis of the immunological relationships between WGA and both *Brachypodium* and rice lectins revealed several interesting features. First, within both genera, all lectins are immunologically indistinguishable

Table 2. Occurrence of lectins in *Gramineae* species.

Subfamily[a] Supertribe Tribe	Genus	Lectin	Specificity	Cross-reaction with WGA-antiserum
Pooideae				
Poanae				
Agrostideae	*Phalaris* (1)[b]	-[c]		-
	Phleum (1)	-		-
Aveneae	*Avena* (3)	-		-
	Arrhenatherum (1)	-		-
Poeae	*Dactylis* (1)	-		-
	Lolium (2)	-		-
	Festuca (2)	-		-
	Poa (2)	-		-
Triticanae				
Brachypodieae	*Brachypodium* (3)	+	NAG, CH[c]	cr[c]
Bromeae	*Bromus* (3)	-		-
Triticeae	cfr. Table 1 (77)	+	NAG, CH	li[c]
Oryzoidea	*Oryza* (20)	+	NAG, CH	cr
Panicoidea				
Andropogoneae	*Sorghum* (2)	-		-
	Zea (1)	-		-
Paniceae	*Panicum* (1)	-		-

a Classification according to Macfarlane, Watson (1982).
b Number of species tested.
c Abbreviations: +: positive reaction; -: no reaction; NAG: N-acetylglucosamine; CH: chitin hydrolysate; cr: cross-reaction; li: line of identity with the precipitin line of WGA upon immunodiffusion against WGA-antiserum

from each other since their respective precipitin lines fuse completely upon double immunodiffusion against WGA-antiserum (Fig. 1). Hence, one can conclude that just as in the case of cereal lectins, *Brachypodium* and rice lectins represent two classes of closely related proteins. Second, both classes of lectins cross-react with WGA-antiserum to a different extent (Fig. 1) and it appears that *Brachypodium* lectins are more closely related to WGA than rice lectins (as the latter form precipitin lines at a higher WGA-antibody concentration than the former).

Because of the obviously close serological relationships between on the one hand WGA, and on the other hand *Brachypodium* and rice lectins, it looked worthwhile to investigate possible

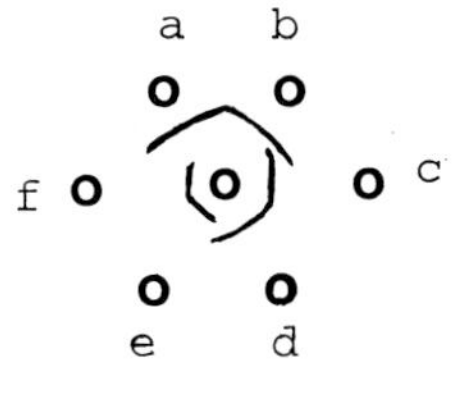

Fig.1: Schematic representation of double immunodiffusion of wheat (a), rye (b), *Brachypodium sylvaticum* (c), *Brachypodium pinnatum* (d), *Oryza sativa* (e) and *Oryza nivara* (f) lectin agaist WGA-antiserum.

structural similarities between these lectins. Therefore, we isolated lectins from embryos of *Brachypodium sylvaticum* and *Oryza sativa* (common rice) and characterized them partially with respect to their biochemical and physico-chemical properties.

The *Brachypodium sylvaticum* lectin (BsyL) is a dimer composed of identical subunits of molecular weight 18,000. The subunits are (not covalently) bound into dimers which dissociate at low pH (below 3.0) (Peumans *et al.* 1982a). Therefore, it appears that BsyL strikingly resembles WGA for what concerns its structural properties. However, since BsyL failed to exchange subunits *in vitro* with cereal lectins (whereas different cereal lectins readily do so with each other), there must be some structural difference(s) between both types of lectins, which is of course in agreement with the above described serological differences between cereal and *Brachypodium* lectins.

Rice lectin (RL) also resembles cereal lectins in that it has the same molecular weight (36,000) and a strikingly similar amino acid composition, but on the other hand it differs in several aspects. Like WGA, RL is composed of 2 identical subunits of molecular weight 18,000. However, unlike WGA these subunits are for the greater part (90%) cleaved in 2 smaller polypeptides of molecular weight 10,000 and 8,000 (Peumans *et al.* 1983). Besides, RL does not dissociate at low pH and has a 4-fold higher specific agglutination activity than cereal lectins. In spite of these obvious differences (which are greater than those between cereal and *Brachypodium* lectins), RL still cross-reacts with WGA-antiserum.

Conclusions

A general conclusion to be drawn from the above described results is that within the *Gramineae* family only one major type of agglutinin occurs. It can be divided further into three subtypes, namely the cereal, *Brachypodium* and rice lectins. Since these three types are both structurally and serologically related to each other, they might have a common evolutionary ancestor. In addition, one can expect that the degree of structural and immunological relationship between these lectins reflects the phylogenetic relationship between the species in which they occur. The finding that within the Triticeae tribe all lectins are serologically identical, and that *Brachypodium* lectins are immunologically related but not identical to WGA, fits with the currently accepted phylogenetic relationship within the Triticeae tribe and Triticanae supertribe. The only exception are the members of the Bromeae tribe which do not contain any lectin or serologically related protein. This is however only true for the most recent classification system where the genus *Bromus* was placed in the Triticanae supertribe (Macfarlane, Watson 1982). In fact, the authors admitted that *Bromus* differed considerably from *Brachypodium* and Triticeae species. According to our results, a chemotaxonomic classification based on lectins would not place the genus *Bromus* in the Triticanae supertribe but in the Poanae supertribe. A more fundamental problem arises when one tries to correlate the serological and structural similarities between cereal and rice lectins with the current classification of these taxonomic groups in two different subfamilies. Since both types of lectins are without any doubt closely related, there might be a much closer taxonomic relationship between rice and cereals than appears from their current classification. An alternative classification in which the chemotaxonomic data on the distribution and the serological relationship of lectins are taken into consideration is given in Fig.2.

Regardless of the exact phylogenetic classification of lectin-containing grass species, the structural similarities between their lectins indicate a high degree of conservation during the evolution of the *Gramineae*. Structural and serological relationships have been studied also for different solanaceous (Kilpatrick *et al.* 1980) and legume lectins (Foriers *et al.*) 1979; Hankins *et al.* 1979). In these cases, however, immunological differences could be observed between species belonging to the same genus.

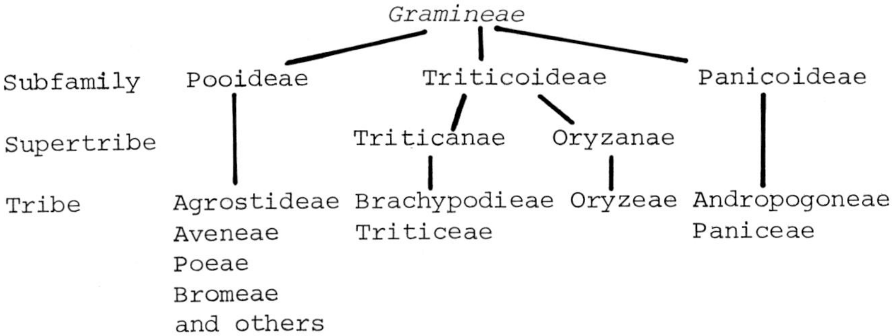

Fig.2: Alternative classification of part of the *Gramineae* based on the occurrence and serological relationship of lectins.

MOLECULAR BIOLOGY OF CEREAL AND RICE LECTINS

One of the major problems we had to deal with during our search for the physiological role of cereal (and related) lectins, was the complete lack of insight in some of the fundamental molecular-biological aspects of their biology. Since no working hypothesis could be formulated without an understanding of how, when and where these proteins are synthesized and possibly processed and transported, we tried to answer these questions by using developing and germinating cereal and rice embryos as model systems.

Lectin Accumulation in Developing and Germinating Wheat Embryos

Since cereal lectins are present in dry embryos, they are probably synthesized during seed formation. To investigate more precisely the time of their synthesis we followed the accumulation of lectin in developing wheat and rye embryos throughout embryogenesis (Peumans *et al.* 1982e). Thereby, it became evident that the lectin content of the embryos increased dramatically during the period coinciding with the development of the primary axes. These were previously found to contain all the lectin present in the embryos (Stinissen *et al.* 1981). Once the development of the organ was completed, the lectin content did not increase further.

Although indicative of a maturation-specific synthesis, these results could not exclude possible additional lectin synthesis during germination. To answer this important question, we followed the new-synthesis of lectin in isolated wheat and rye embryos during early germination. It was observed that these embryos do synthesize lectin during the initial stage of germination. However, on the basis of kinetical analyses and the results of mRNA inhibition experiments with cordycepin as the inhibitor, it could be concluded that this lectin synthesis is exclusively mediated by long-lived mRNAs. Since these mRNAs disappear with a half-life of about 3 h, lectin synthesis falls back under the level of detection within 24 h. Hence, it can be concluded that cereal lectins are typical maturation-specific proteins, which, however, does not preclude that even 34-day-old plants still contain about 50% of the original lectin content of the embryo (Mishkind *et al.* 1981).

Lectin Synthesis and Accumulation in Developing and Germinating Rice Embryos

Since rice lectin resembles cereal lectins in many aspects, it could be expected that this agglutinin too is specifically synthesized during embryogenesis. In order to confirm this logical supposition, we followed the accumulation of lectin in maturing rice grains and tried to correlate it with the development of the embryos.

As shown in Fig.3, rice lectin could be detected as early as 8 days post anthesis (DPA). From then on, the lectin content of the embryos increased dramatically for about 8 days, after which the increase slowed down progressively until a final lectin content was reached around 21 DPA. Again, the accumulation of the lectin appears to coincide with the phase of rapid embryo development.

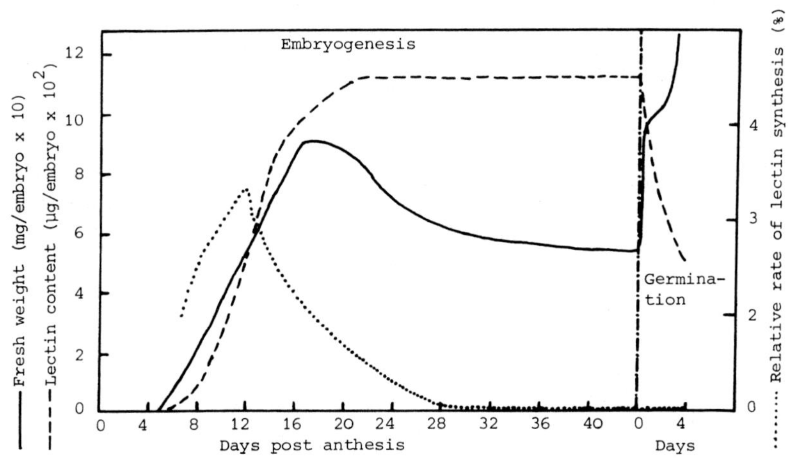

Fig.3: Embryo development, lectin accumulation and relative rate of lectin synthesis in rice embryos during embryogenesis and early germination (of isolated embryos).

To find out whether the dramatic increase in lectin content between 8 and 16 DPA resulted from an enhanced rate of lectin synthesis, embryos were excised at different developmental stages and labelled with ^{35}S cysteine. Subsequently, extracts were made from the labelled embryos and the ^{35}S cysteine incorporation in both total protein and affinity-purified lectin determined. From these data, the relative rate of lectin synthesis (defined as the % of total incorporated ^{35}S cysteine which was recovered in the lectin fractions) was calculated. As shown in Fig.3, the relative rate of lectin synthesis is high at the time of rapid lectin accumulation and decreases dramatically once the accumulation rate slows down. Since embryos from desiccating grains (24 DPA) synthesize very little lectin, whereas their total protein synthesis activity is as high as that of younger embryos, it is evident that rice lectin can be considered as a typical maturation-specific protein. This is confirmed by the finding that embryos isolated from fully mature rice grains do not synthesize lectin during early germination, or at least not in detectable amounts. Although at present the fate of the embryo lectin is unknown, it is worthwhile to mention here that, at least isolated embryos, loose half of their lectin content within 3 days of germination.

A final remark to be made here is that the rice lectin is exclusively synthesized in the primary axes (which of course is in agreement with the localization of the lectin). However, because of practical reasons the above described experiments were performed with intact embryos. It is evident that because of the protein synthesis in the scutella, the relative rate of lectin synthesis in the embryos is lower than that in the primary axes. However, since this relative rate represents only an arbitrary and relative measure, the conclusions about changes in lectin synthesis as a function of seed development remain valid.

Hormonal Control of Lectin Synthesis in Developing Rice Embryos.

In a recent report on the control of WGA synthesis in developing wheat embryos, evidence was presented that this lectin is a typical dormancy-specific protein whose synthesis is maintained by abscissic acid (ABA) (Triplett, Quatrano 1982). To find out whether phytohormones also play a regulatory role in developing rice embryos, lectin synthesis was followed in isolated 12 DPA embryos grown on a culture medium in the presence and absence of ABA and gibberellic acid (GA_3). As shown in Fig.4, the synthesis of lectin is strongly promoted by ABA whereas it is almost completely suppressed by GA_3. When both ABA and GA_3 are present, lectin synthesis is comparable to that in ABA embryos which indicates that the promoting effect of ABA (on lectin synthesis) is not counteracted by GA_3. Since the embryos do not develop in the presence of ABA, whereas the control and GA_3-treated embryos form roots and shoots within 3 days, there is little doubt

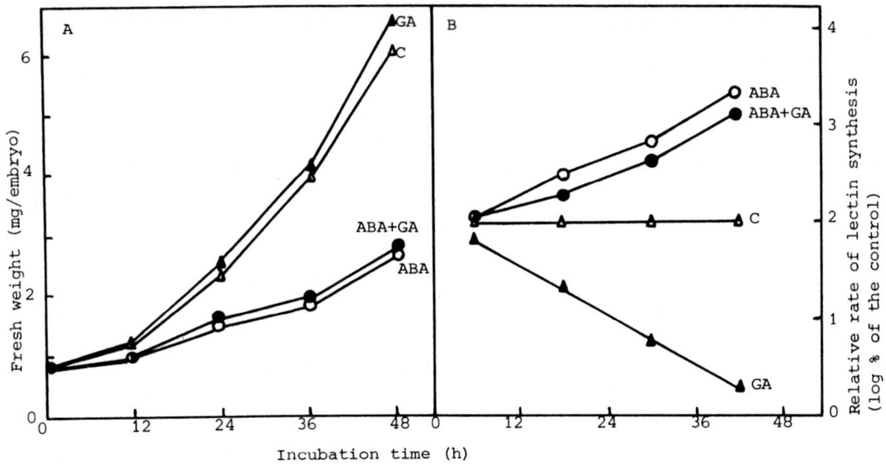

Fig.4: Effect of ABA (25 μM), GA (50 μM) and GA+ABA (50 and 25 μM respectively) on:
A. The development of 14 DPA rice embryos in culture medium.
B. The relative rate of lectin synthesis in the same embryos. This rate of synthesis was determined over the 12 final h of incubation in the culture medium (after preincubation for 12, 24 and 36 h respectively). Results are expressed as % of the control (no GA or ABA added) on a logarithmic scale.

that rice lectin is a dormancy-specific protein that is specifically synthesized upon the entry of the embryos into their natural or exogenously induced developmental arrest.

Synthesis and Processing of Cereal and Rice Lectins

Having established where and when cereal and rice lectins are synthesized and how their synthesis is regulated, some fundamental molecular processes such as their synthesis and possible processing and transport remained to be elucidated. Using *in vivo* labelling and pulse-chase labelling experiments, the sequence of events leading from the primary translation products to the final lectin polypeptides could be traced both in germinating wheat and rye embryos and in developing rice embryos (Stinissen *et al.* 1982; Stinissen *et al.* 1983b).

As indicated in the scheme shown in Fig.5, both cereal and rice lectins are synthesized *in vivo* as high-molecular-weight precursors (MW 23,000) which are post-translationally processed in a single step into 18,000 MW polypeptides. In

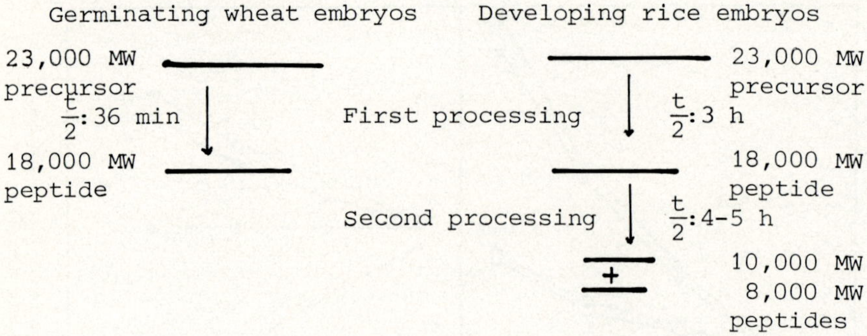

Fig.5: Schematic representation of *in vivo* synthesis and processing of lectins in germinating wheat and in developing rice embryos.

the case of cereal lectins, the 18,000 MW polypeptide is the end product, whereas in rice it is further cleaved (but incompletely) in a second step into smaller polypeptides of MW 10,000 and 8,000 respectively. Kinetic analyses of the processing in both systems allowed to determine the half-life of the respective 23,000 MW precursors. The apparent discrepancy between the half-life of the precursor in cereals (36 min) and rice (3 h) has to be ascribed to the fact that in the former case germinating embryos, and in the latter maturing embryos were used (since in developing wheat embryos the half-life of the precursor is about 3 h).

We also followed the synthesis and processing of cereal and rice lectins *in vitro*, both in a homologous system (derived from cereal and rice embryos respectively) and in *Xenopus* oocytes. In the latter system, only the lectin precursors could be detected among the translation products of injected wheat or rice embryo mRNA (Fig.6), indicating that (1) the translation product of the lectin mRNA is indeed a 23,000 MW precursor, and (2) the oocytes cannot perform the processing of this precursor. The homologous systems, on the contrary, do not contain any precursor among the translation products of their endogenous lectin mRNA. Indeed, as indicated in Fig.6, the only lectin product detected in a wheat or rye embryo cell-free system is the 18,000 MW polypeptide, whereas in a rice system both the 18,000 MW polypeptide and the small 8,000 and 10,000 MW polypeptides are recovered. It appears, therefore, that the homologous cell-free systems are fully capable of performing the processing step(s) which usually occur *in vivo*. Moreover, since even after short incubation times (less than 5 min), no precursor can be detected, the

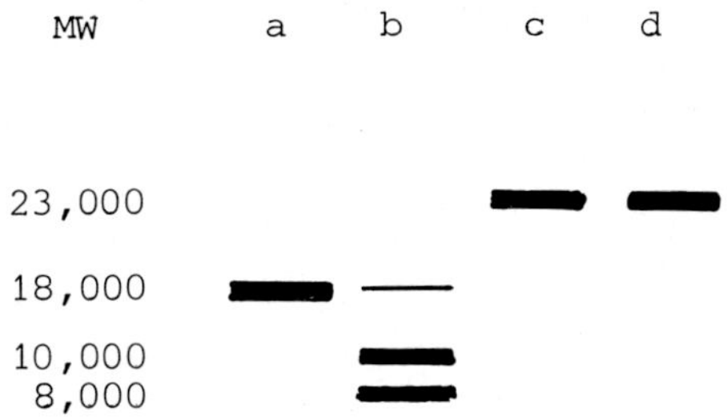

Fig.6: Schematic representation of lectin products synthesized *in vitro* in homologous extracts from wheat (a) and rice (b) embryos, and in *Xenopus* oocytes programmed with wheat (c) and rice (d) embryo mRNA.

processing occurs at a much higher rate *in vitro* than *in vivo*, which indicates that *in vivo* precursor and processing enzyme(s) are located in different compartments.

A major conclusion to be drawn from these observations is that the processing of the 23,000 MW lectin precursors is essentially a post-translational event, and thus differs fundamentally from the co-translational processing of the signal peptide-carrying precursors of legume and castor bean lectins (Hemperly *et al.* 1982; Higgins *et al.* 1982; Roberts, Lord 1981). On the basis of the striking resemblances between the processing of the 23,000 MW lectin precursors and the so-called 'vectorial processing' of cytoplasmically made chloroplast, mitochondrial and glyoxysomal proteins (Smith, Ellis 1979; Maccecchini *et al.* 1979; Gietl, Hock 1982), it seems likely that both types of post-translation protein modifications are similar and probably have an analogous function. If so, one could expect that cereal and rice lectins are destined to be transported across a membrane system into an organelle or an extra-cytoplasmic compartment. Unfortunately, it is not clear which organelles or compartment is involved. In our opinion, the vectorial processing of cereal and rice lectins is a means of transporting these proteins through the cytoplasmic membrane into the periplasm and/or cell wall. This would be in agreement also with the immunocytochemical localization of part of the lectin in imbibed embryos at the protoplast-cell wall interface (Mishkind *et al.* 1982).

PHYSIOLOGICAL ROLE OF CEREAL AND RELATED LECTINS

One of the first well-defined hypotheses concerning the physiological role of plant lectins has been advanced for WGA. On the basis of the carbohydrate-binding specificity of this lectin and its inhibitory effect on fungal growth,

Mirelman *et al.* (1975) put forward the hypothesis that WGA possibly protects germinating seeds against fungal attack. Although in a recent paper dealing with the immunocytochemical localization of WGA (Mishkind *et al.* 1982), the association of WGA with cells and tissues that establish direct contact with the soil during germination and growth of the plant has been interpreted in favour of a defensive role, several serious objections can be made against this proposed function.
- it is questionable whether WGA concentrations under natural germination circumstances can be high enough to exert any protective function.
- if the lectin has a role to fulfill in the germinating seed, why should it then be synthesized during early embryogenesis and not during germination, why should it then be a dormancy-specific protein?
- since cereals are, even in a young state, susceptible to numerous leaf pathogens, why is there no lectin in the leaves?
- wild relatives of modern cereals are more resistant against a number of pathogenic fungi, nevertheless their embryos contain much less lectin than these of cultivated varieties.
- to what extent can the results obtained with modern wheat be extrapolated to its wild relatives in which the lectins evolved? It should not be forgotten that these primitive cereals have a completely different biology (e.g. their seeds are firmly encapsulated in spikelets and for a long time exposed to environmental conditions in a dormant state).

As an alternative to the defensive role of WGA, we propose a specific and endogenous physiological function for cereal and related lectins. For that purpose all evidence available in favour of such a role is compiled and together with the most recent findings concerning the molecular-biology of cereal lectins, integrated into a working hypothesis of what lectins do and how they act.

Evidence for a Specific Role of Cereal Lectins

Several lines of indirect evidence indicate that cereal and related lectins fulfill a specific function.
1. Cereal, *Brachypodium* and rice lectins represent a large group of closely related proteins, which obviously have been very well conserved during the evolution and divergency of the grass family.
2. In a screening experiment of more than 1,000 varieties and lines of wheat, rye, barley, rice and some of their wild relatives, we could not detect any lectin-linus line. This

observation suggests that (1) *Gramineae* lectins are required for normal development at one or another stage of the life cycle of the plant, and (2) that in order to ensure their proper function, the molecular structure of these lectins has been conserved rigorously.

Where and When Do *Gramineae* Lectins Function?

Speculations about the role of cereal lectins have to take into consideration the results of recent studies concerning some of their botanical and molecular-biological aspects, since these give valuable information about where and when these lectins may act. Indeed, from the data discussed in the second section, it can be concluded that both cereal and rice lectins are exclusively synthesized in the primary axes during embryogenesis. Moreover, cereal and rice lectins are not only maturation-specific proteins, but also appear to be dormancy-specific,since they are synthesized during the natural or ABA-induced entry into developmental arrest of the embryos. It is obvious, therefore, that these lectins play a role in the developing primary axes, and that their function is somehow related to the establishment of the resting state in this organ.

Do Cereal and Rice Lectins Control the Development of the Primary Axes During Embryogenesis?

One of the major problems in the physiology and biochemistry of seed formation is the regulation of a number of processes which prevent precocious 'germination' of the primary axis. Indeed, whereas cells of specialized tissues (such as cotyledons and endosperm) are fully differentiated and not able to divide anymore, the primary axis represents a miniature plant, composed merely of meristematic and undifferentiated cells. Since the primary axis is already in an early developmental stage fully able to develop into a seedling when put into the appropriate medium, the maturing seed has to be provided with a control system that prevents precocious germination. Naturally, such a system must be reversible since otherwise germination would be impossible. In addition, it must be specific enough to allow other seed parts to complete their development.

Although it is generally believed that plant growth regulators play a key role in the control of seed formation, other factors may be involved as well. In our opinion, cereal and related lectins represent a system, or at least part of a system, that keeps developing embryonic axes in a resting state. Such a function would explain why these lectins are exclusively synthesized in developing primary axes and behave as dormancy-specific proteins. In addition, the specificity and reversibility of the carbohydrate binding of these lectins allows them to establish a highly specific and fully reversible control mechanism. Finally, the physiological role proposed here would account for the high degree of evolutionary conservation of cereal and related lectins and the apparent lack of lectin-minus lines.

Working Mechanism of Cereal and Related Lectins

The maintenance of a resting state in developing embryonic axes requires a stringent control of cell division in their meristems. Indeed, whereas in tissues such as endosperm, scutellum and aleuron layer, the mitotic activity becomes completely and irreversibly blocked in an early stage of seed formation, it is only temporarily arrested in developing primary axes (as they resume cell divisions shortly after the beginning of imbibition). This means that there is a mechanism that inhibits - in a completely reversible way - mitotic activity during a great part of embryonic axis development. Cereal and related lectins are good candidates for this function because of their specific and reversible carbohydrate-binding activity. The exact working mechanism remains unknown, but it seems likely that the lectins act by binding to receptor molecules at the outer side of the cytoplasmic membrane (which could explain the localization of at least a part of the lectin at the protoplast-cell wall-interface).

Summarizing, it can be concluded that lectins may have a regulatory function in the embryogenesis of cereals and related species. Conclusive evidence for such a specific endogenous role is still scarce, but it is surely worthwhile to look further in this direction.

Acknowledgements. W.P. is Research Associate of the National Fund for Scientific Research (Belgium). H.S. receives a fellowship of the I.W.O.N.L. (Belgium).

REFERENCES

Allen AK, Neuberger A, Sharon N (1973). The purification, composition and specificity of Wheat Germ Agglutinin. Biochem J 131: 155.

Foriers A, De Neve R, Kanarek L (1979). Lectin sequences as a tool for chemotaxonomical classification. Physiol Vég 17:597.

Gietl C, Hock B (1982).Organelle-bound matate dehydrogenase isoenzymes are synthesized as higher molecular weight precursors. Plant Physiol 70:483

Hankins CN, Kindinger JI, Shannon LM (1979). Legume lectins. I Immunological cross-reactions between the enzymic lectin from mung beans and other well characterized legume lectins. Plant Physiol 64:104

Hemperly JJ, Mostov KE, Cunningham BA (1982). *In vitro* translation and processing of a precursor form of favin, a lectin from *Vicia faba*. J Biol Chem 257;7903

Higgins TJV, Chandler PM, Spencer D, Chrispeels MJ, Zurawski G (1982). The synthesis, processing and primary structure of pea seed lectin. NATO FEBS Advanced Course Porto Portese Italy. Abstract p 9.

Kilpatrick DC, Jeffree CE, Lockhart CM, Yeoman MM (1980). Immunological evidence for structural similarity among lectins from species of the *Solanaceae*. FEBS Lett 113:129

Maccecchini ML, Rudin Y, Blobel G, Schatz G (1979). Import of proteins into mitochondria: precursor forms of the extramitochondrially made F_1-ATPase subunits in yeast. Proc Natl Acad Sci USA 76:343

Macfarlane TD, Watson L (1982). The classification of Poaceae subfamily Pooideae. Taxon 31:178

Mirelman D, Galun E, Sharon N, Lotan R (1975). Inhibition of fungal growth by Wheat Germ Agglutinin. Nature 256:414.

Mishkind M, Keegstra K, Palevitz BA (1980). Distribution of Wheat Germ Agglutinin in young wheat plants. Plant Physiol 66:950.

Mishkind M, Raihkel NV, Palevitz BA, Keegstra K (1982). Immunocytochemical localization of Wheat Germ Aggluninin in wheat. J Cell Biol 192:753.

Nagata Y, Burger MM (1972). Wheat Germ Agglutinin. Isolation and Characterization. J Biol Chem 247:2248.

Nagata,Y, Burger MM (1974). Wheat Germ Agglutinin. Molecular characteristics and specificity for sugar binding. J Biol Chem 249:3116.

Peumans WJ, Spaepen C, Stinissen HM, Carlier AR (1982a). Isolation and partial characterization of a lectin from a false brome grass (*Brachypodium sylvaticum*). Biochem J 205:635.

Peumans WJ, Stinissen HM, Carlier AR (1982b). Isolation and partial characterization of WGA-like lectins from rye (*Secale cereale*) and barley (*Hordeum vulgare*) embryos. Biochem J 203:239.

Peumans WJ, Stinissen HM, Carlier AR (1982c). A genetic basis for the origin of six different isolectins in hexaploid wheat. Planta 154:562

Peumans WJ, Stinissen HM, Carlier AR (1982d). Subunit exchange between lectins from different cereal species. Planta 154:568

Peumans WJ, Stinissen HM, Carlier AR (1982e). Lectin synthesis in developing and germinating wheat and rye embryos. Planta 156:41

Peumans WJ, Stinissen HM, Tierens M, Carlier AR (1982f). *In vitro* synthesis of lectins in cell-free extracts from dry wheat and rye embryos. Plant Cell Rep 1:212

Peumans WJ, Stinissen HM, Carlier AR (1983). The rice lectin and its relationship to cereal lectins. Biochem Physiol Pflanz In the press

Rice RH (1976). Wheat germ agglutinin. Evidence for a genetic basis for multiple forms. Biochim Biophys Acta 444:175

Rice RH, Etzler ME (1974). Subunit structure of wheat germ agglutinin. Biochim Biophys Res Commun 59:414

Roberts LM, Lord JM (1981). The synthesis of *Ricinus communis* agglutinin. Cotranslational and post-translational modification of agglutinin polypeptides. Eur J Biochem 119:31

Smith SM, Ellis RJ (1979). Processing of small subunit precursor of ribulose biphosphate carboxylase and its assembly into whole enzyme are stromal events. Nature 278:662

Stinissen HM, Peumans WJ, Carlier AR (1981). Some biological aspects of the wheat embryo lectin. Arch Int Physiol Biochim 89:132

Stinissen HM, Peumans WJ, Carlier AR (1982). *In vivo* synthesis and processing of cereal lectins. Plant Molec Biol 1:277

Stinissen HM, Peumans WJ, Carlier AR (1983a) Occurrence and immunological relationships of lectins in gramineous species. Planta In the press

Stinissen HM, Peumans WJ, Carlier AR (1983b) Two-step processing of *in vivo* synthesized rice lectin. Plant Molec Biol In the press

Triplett BA, Quatrano RS (1982). Timing, localization and control of wheat germ agglutinin synthesis in developing wheat embryos. Devl Biol 91:491

Wright CS (1980) Crystallographic elucidation of the saccharide binding mode in wheat germ agglutinin and its biological significance. J Mol Biol 141:267

Chemical Taxonomy, Molecular Biology, and Function of Plant Lectins, pages 117–130

CONTROL OF THE SYNTHESIS AND LOCALIZATION OF WHEAT GERM AGGLUTININ DURING EMBRYOGENESIS[1]

Ralph S. Quatrano, Roswitha Hopkins and Natasha V. Raikhel[2]

Department of Botany and Plant Pathology, Oregon State University, Corvallis, OR 97331

We are interested in understanding the physiological and molecular controls of embryogenesis in higher plants. Wheat embryos in culture provide a model system to experimentally approach the regulatory events associated with the process of embryogenesis. Two developmental pathways exhibited by most higher plant embryos can be dissociated when immature wheat embryos are removed from the grain and cultured in the presence or absence of the growth regulator abscisic acid (ABA) (Triplett, Quatrano 1982). In the presence of ABA (+ABA) wheat embryos in culture continue to undergo elaborate tissue and molecular differentiation, similar to embryos remaining on the plant (maturation pathway). A number of other studies have also reported that ABA inhibits germination of embryos in culture while promoting embryogenesis (Crouch, Sussex 1982; Ammirato 1983; Umbeck, Norstog 1979; Dunwell 1981). In the absence of ABA (-ABA), the maturation pathway is abbreviated or absent and the embryos in culture precociously germinate (germination pathway). Changes in the level of endogenous ABA in developing wheat grains are consistent with its role as a regulator *in vitro*, i.e. ABA is found in high concentrations at those stages of wheat grain development when the embryo is capable of germinating (King 1976). Thus,

[1]This work was supported by a grant from the U.S.D.A. Competitive Grants Program (59-2411-0-1-4110) to R. S. Q.
[2]Department of Botany, University of Georgia, Athens, GA 30602.

we have defined an experimental system in which wheat embryos in culture can express both morphologically and molecularly the maturation or germination pathway of development (see Figure 1).

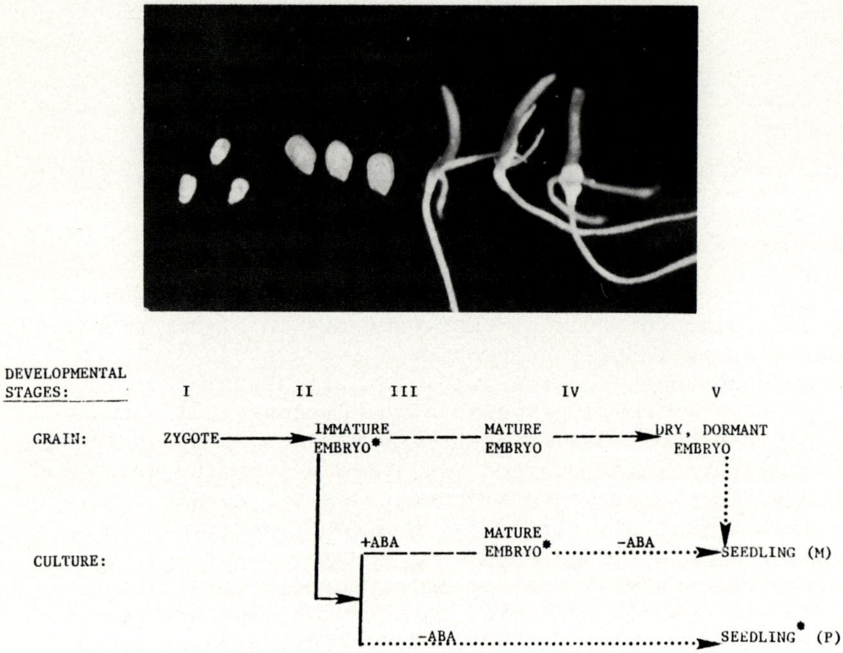

Fig. 1. Diagram showing the various developmental pathways of the wheat embryo in the grain and isolated in culture. The maturation pathway is shown as dashed lines while the germination pathway is shown as dotted lines. Seedlings formed from embryos in culture by precocious germination are referred to seedlings (P), and distinguished from seedlings formed from embryos that have proceeded through the maturation pathway, seedlings (M). Asterisks refer to stages pictured in the photograph: Stage III wheat embryos at the beginning of incubation (left), after 5 days in culture +ABA (middle) and after 5 days in culture -ABA (right). Note embryo growth and scutellum enlargement in +ABA embryos. Photograph from (Triplett, Quatrano 1982) Courtesy Academic Press.

Molecular markers unique to each pathway have been identified as a start towards understanding the genetic control mechanisms involved with regulation of wheat embryogenesis. We recently reported that scutellar storage proteins are associated with the maturation pathway, while the small subunit of Rubp carboxylase is unique to the germination pathway (Quatrano, Ballo, Williamson, Hamblin, Mansfield 1983). We also isolated several cDNA clones whose sequences code for proteins that are expressed during maturation. Evidence was presented showing that ABA prevents expression of the germination marker, carboxylase, at the level of transcription (Quatrano, Ballo, Williamson, Hamblin, Mansfield 1983). This paper will present evidence that wheat germ agglutinin (WGA) is an embryo-specific protein, a member of the maturation gene set, and whose synthesis and localization during grain development appears to be under ABA control.

MATERIALS AND METHODS

Wheat plants (Triticum aestivum, L. var. Chinese Spring and Yamhill) were grown and grain developmentally staged as reported earlier (Rogers, Quatrano 1983). Embryos were isolated and cultured in the dark at 27°C on filter paper containing a liquid medium (Triplett, Quatrano 1982) and labeled with carrier-free $H_2{}^{35}SO_4$ (New England Nuclear-43 Ci/mg) at 30 μCi/ml in the absence or presence of 10^{-4} M ABA or 5% (w/v) mannitol. WGA was extracted from embryos by homogenization in 50 mM HCl, stirred for 30 minutes at 4°C and then centrifuged (12,000 g - 10 min). The supernatant was dialyzed (Spectrapor 3) against 10 mM ammonium carbonate (2 changes) for 2 hours at 4°C, centrifuged and lyophilized. Antibody to WGA was prepared (Triplett, Quatrano 1982) or purchased from Vector Laboratories, Inc. (Burlingame, CA 94010). Staphylococcus aureus cells (IgGsorb - The Enzyme Center, Boston, MA 02111) were used for immunoprecipitation (Martial, Baxter, Goodman, Seeburg 1977). WGA was assayed also by a chitin-binding assay. Washed chitin (50 mg) was incubated for 30 min in a 0.5 ml microfuge tube with a ^{35}S labeled embryo extract (3.5×10^6 cpm) in 10 mM Tris-HCl (pH 8.5). The chitin was then washed (by vortexing, and centrifuging in a Beckman microfuge) with 0.3 ml Tris buffer (2 X), buffer containing 1 M NaCl (4 X) and finally with buffer alone (6 X). WGA was released from chitin with several washes of 50 mM HCl. The washes were pooled and lyophilized. Proteins were electrophoresed in polyacrylamide gels using a sodium dodecyl sulfate (SDS) system (Thomas, Kornberg 1975), or by isoelectric focusing (10°C) on LKB Ultro PAG plates (pH 3.5-9.5) at constant power (25 W), 2000 V/50 mA for 60 min. RNA was isolated using a LiCl extraction procedure (Hall, Ma, Buchbinder, Pyne, Sun, Bliss 1978) and Poly A RNA was prepared by elution from Poly U Sepharose and translated in a rabbit reticulocyte system (Bethesda Research Laboratories, Gaithersburg, MD) using ^{35}S methionine. For fluorography, gels were soaked in salicylate (Chamberlain 1979), dried, placed under Kodak film (XAR-5) and stored at -70°C. Immunocytochemistry was performed on embryos fixed in 4% formaldehyde (4°C-24 hrs) and stored in PBS (Triplett, Quatrano 1982) at 4°C. WGA was localized in frozen sections of fixed embryos by the peroxidase/antiperoxidase procedure using anti-WGA (Sternberger 1979; Mishkind, Raikhel, Palevitz, Keegstra 1982). Sections were viewed with a Reichert Zetopan microscope equipped with differential interference contrast optics.

RESULTS AND DISCUSSION

Effect of ABA on Embryonic Proteins

The acid-soluble proteins that accumulate in Stage III embryos (var. Chinese Spring) cultured for 3 days in +ABA or -ABA are shown in Figure 2. The acid fraction comprises about 50% of the total protein complement and is enriched in +ABA embryos (Triplett, Quatrano 1982). When the +ABA extract is subjected to isoelectric focusing, approximately 50 protein species are discernible (Fig. 3). These proteins that accumulate in +ABA embryos in culture are identical to the proteins found in Stage V embryos that have matured on the plant (Quatrano, Ballo, Williamson, Hamblin, Mansfield 1983). Hence, ABA stimulates the precocious synthesis of maturation-specific proteins in Stage III embryos in culture.

Stage V embryos incubated in culture with +ABA and -ABA exhibit similar morphological and molecular responses as Stage III embryos. Only quantitative differences are evident between the proteins present in Stage III and V embryos after incubation in +ABA (Fig. 3). Apparently, Stage V embryos in +ABA continue to accumulate in culture, those

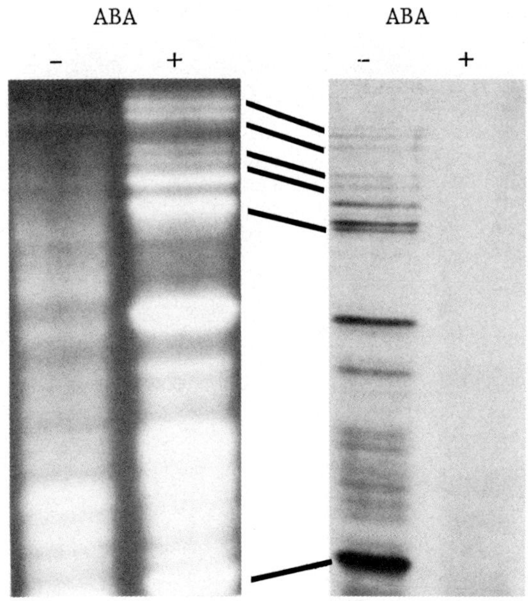

Fig. 2. SDS gel separation of ^{35}S labeled acid soluble proteins extracted from embryos cultured in +ABA and -ABA. Right panel is a Coomassie Blue-stained gel while the left panel is an fluorograph of the same extract. The upper group of proteins have molecular weights between 50-75,000, the prominent band is the middle of the gel is about 35,000, and the distinct band at the bottom of the gel is approximately 12,000. The same number of embryos was used for each extract.

proteins that were synthesized during Stages III and IV on the plant. Therefore, embryos at different developmental stages can respond to ABA in culture by the synthesis of the same proteins which are characteristic of the maturation pathway.

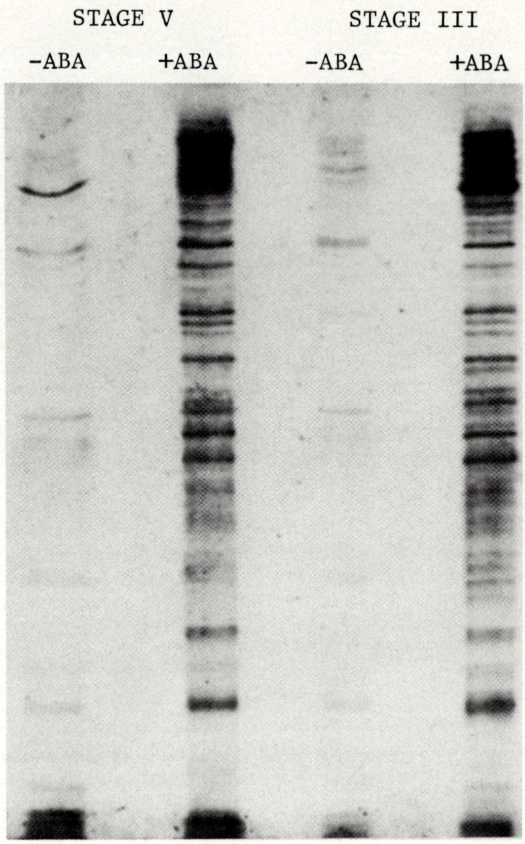

Fig. 3. Isoelectric focusing of acid soluble proteins extracted from Stage III and Stage V embryos cultured in +ABA and -ABA. The pH gradient in the gel stained in Coomassie Blue is from 3.5 (top) to 9.5 (bottom). The same number of embryos was used for each extract.

Figure 3 also shows that most of the maturation-specific proteins (+ABA) are absent from 3-day-old seedlings (-ABA). Stage III embryos that precociously germinate into seedlings (seedling (P)-Fig. 1), bypass the maturation pathway and never synthesize the maturation proteins. Stage V embryos that germinate to form seedlings (seedlings (M)-Fig. 1) contained the maturation-specific proteins at the start of the -ABA incubation but these apparently are lost during germination (Fig. 3). The absence of the majority of the prevalent embryonic proteins in seedlings (M) from Stage V embryos is consistent with their preliminary identification as scutellar storage proteins (Quatrano, Ballo, Williamson, Hamblin, Mansfield 1983). However, several acid-soluble proteins synthesized during the maturation pathway are present in 3-day-old seedlings (Fig. 3 - Stage V, -ABA).

Effect of ABA on WGA Synthesis

A protein found in the acid-soluble fraction is WGA. Previous work of ours (Triplett, Quatrano 1982) confirmed an earlier report (Mishkind, Keegstra, Palevitz 1980) that WGA is an embryo-specific protein. In addition, WGA appears in the embryo during Stages III and IV of grain development on the plant (Table 1). Given the facts that WGA is embryo-specific and accumulates during maturation, we asked if Stage II/III embryos that bypass the maturation pathway (i.e. cultured in -ABA) accumulate WGA. Table 2 shows that in -ABA, Stage III embryos precociously germinate and do not accumulate normal levels of WGA. In contrast, embryos cultured in +ABA for 5 days accumulate about 1/2 the amount of WGA found in mature Stage V embryos. Hence, the accumulation of WGA appears to be under the control of ABA and regulated like other members of the maturation protein set.

ABA control of the synthesis of WGA can also be demonstrated by incubating embryos with ^{35}S. Embryos incubated in +ABA for 4 days, the last 2 days in the presence of ^{35}S, synthesize WGA as detected by chitin-binding and immunoprecipitation (Fig. 4). Under these conditions, only a trace of WGA synthesis is detected in -ABA embryos.

The synthesis of WGA can also be regulated by the concentration of osmoticum in the medium. Crouch and Sussex (1981) demonstrated in _Brassica_ that embryogenesis and the synthesis of storage proteins in culture can be promoted by high levels of sucrose or mannitol. Such treatment most likely causes water stress in the tissues, which in some cases, is followed by an increase in endogenous levels of ABA (Imamura, Harada 1980). Mannitol treatment of Stage III Chinese Spring embryos results in the synthesis of WGA, similar to the response of embryos

TABLE 1
WGA IN DEVELOPING WHEAT GRAINS[a]

Stage[b]	Tissue	Hemagglutination assay: Protein (mg/ml)	Hemagglutination assay: Activity	Radioimmunoassay (% of protein as WGA)
II	Whole grains	3.7	0	<0.03
III	Embryo	2.2	++	0.20
	Endosperm	1.5	0	<0.03
IV	Embryo	5.7	+++	0.80
	Endosperm	2.8	0	<0.01
V	Embryo	5.8	++++	--
	Endosperm	2.8	0	--
V	Embryo	5.2	++	1.30
	Endosperm	2.7	0	<0.01

[a]Data from Table 2 in (Triplett, Quatrano 1982): Courtesy Academic Press.

[b]Stages according to (Rogers, Quatrano 1983): Refer to Fig 1 of this paper.

TABLE 2
EFFECT OF ABA ON WGA ACCUMULATION IN ISOLATED EMBRYOS[a]

Treatment	Stage[b]	Total acid-soluble protein (ug/embryo)	Radioimmunoassay (% of protein as WGA)
-	Immature embryo	1.12	0.03
-ABA	Seedling (P)	4.22	0.01-0.05
+ABA	Mature embryo	5.68	0.75-0.90

[a]Data from Table 1 in (Triplett, Quatrano 1982): Courtesy Academic Press.

[b]Refer to asterisks in diagram of Fig 1 of this paper.

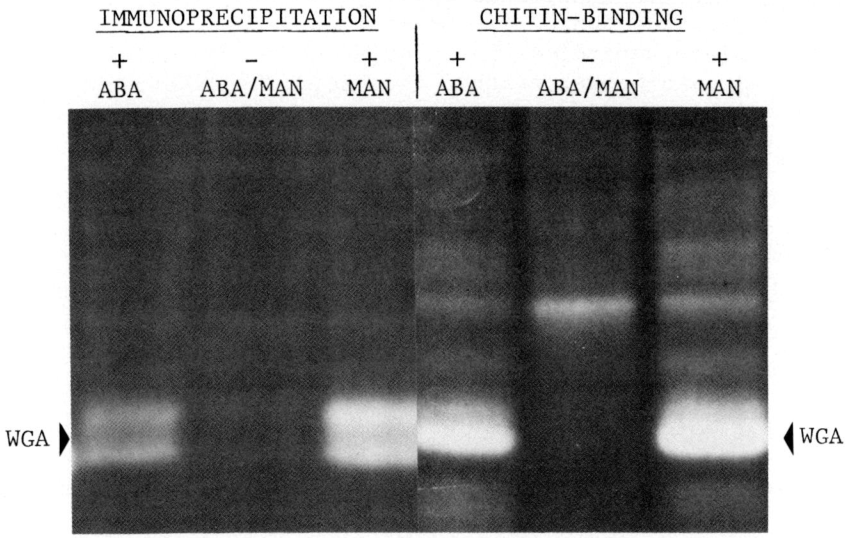

Fig. 4. Fluorogram of a SDS gel separation of ^{35}S labeled WGA from Stage III Chinese Spring embryos treated with ABA or mannitol (MAN). Before electrophoresis, WGA was extracted from embryos with acid and assayed by immunoprecipitation (left panel) or chitin-binding (right panel). Each extract used in the immunoprecipitation or chitin-binding assay contained the same amount of radioactivity.

to +ABA (Fig. 4). It is interesting to note that in both +ABA and +mannitol, two radioactive bands are detectable in the anti-WGA immunoprecipitates (Fig. 4). Also, embryos cultured in +ABA and +mannitol do not significantly increase WGA synthesis over embryos cultured in +ABA or +mannitol alone (data not shown). We are presently measuring ABA levels in +mannitol embryos to determine if the mechanism of mannitol control is via regulation of ABA levels. Neither the ABA concentration in developing wheat embryos nor the osmotic potential of the surrounding endosperm is known, but our results and others (Crouch, Sussex 1981) suggest that both may play an important regulatory role *in vivo*.

In order to understand the level at which ABA influences the expression of the maturation gene set, our goal is to isolate several representative genes including WGA. Since WGA is a member of this gene set, we are attempting to isolate its mRNA directly. WGA synthesis is detected in vitro using a cell-free extract from dry wheat embryos (Peumans, Stinissen, Tierens, Carlier 1982). Previous results of ours indicated that translation products of mRNA's stored in Stage V embryos are electrophoretically the same as some proteins synthesized in Stage III embryos +ABA (Quatrano, Ballo, Williamson, Hamblin, Mansfield 1983). When Poly A RNA from Stage V embryos is translated in a rabbit reticulocyte system, a chitin-binding protein is detected with the same mobility in a SDS gel as WGA (Fig. 5). We have no direct evidence that any of the higher molecular weight products that bind to chitin are WGA precursors. Nevertheless, judging by the amount of a WGA-like protein synthesized in vitro, the level of WGA mRNA does not appear to be high in this fraction. However, other workers (Peumans, Stinissen, Tierens, Carlier 1982) have reported that lectins are synthesized in vitro only in the absence of sulfhydryl compounds. Therefore, our low yields of WGA synthesized in vitro may not be a reflection of the concentration of WGA mRNA. Similar RNA from +ABA

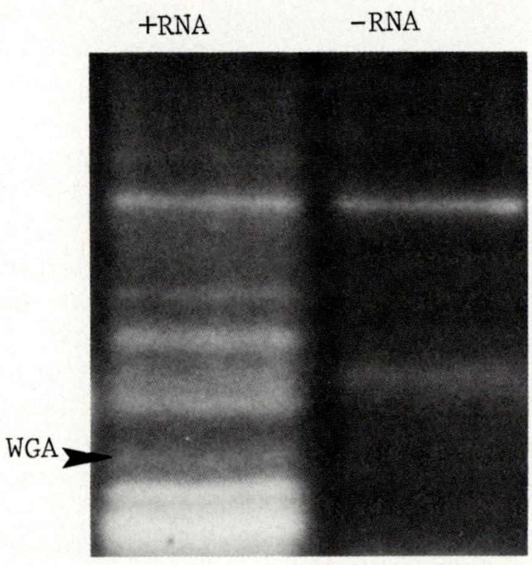

Fig. 5. Fluorogram of a SDS gel separation of chitin-binding proteins synthesized in a cell-free translation system using Poly A RNA from Stage V Yamhill embryos.

Stage III and V embryos is now being translated in both the reticulocyte system and the wheat germ system without sulfhydryl compounds. The levels of mRNA for WGA may be higher in these embryos and the translation efficiency greater in the modified translation system.

As a complementary approach, we have identified several cDNA clones whose sequences code for maturation proteins. These clones were isolated from a cDNA library constructed using Poly A RNA from Stage V embryos (Quatrano, Ballo, Williamson, Hamblin, Mansfield 1983). The probes used were cDNA's prepared from Poly A RNA from Stage III embryos cultured in +ABA and -ABA. The clones isolated represent sequences that hybridized strongly to cDNA from +ABA embryos and not to cDNA from -ABA embryos (Quatrano, Ballo, Williamson, Hamblin, Mansfield 1983). We do not know the protein products of these sequences, but it is possible that one of these might be WGA. These clones are presently being characterized by restriction enzymes and by hybrid-select translation.

Effect of ABA on WGA Localization

WGA is localized in the surface layers of the radicle, adventitious roots, coleoptile and scutellum of Stage V embryos. In adult plants, WGA is found only in the caps and tips of adventitious roots and absent from the aerial parts (Mishkind, Raikhel, Palevitz, Keegstra 1982; Mishkind, Keegstra, Palevitz 1980). Seedlings formed from Stage III embryos which were precociously germinated contained very low levels of WGA (see Table 2). What is the distribution of WGA in seedlings (P) that bypass the maturation pathway? To our surprise, we found immunochemically detectable WGA only in the radicle and absent from the coleoptile of -ABA embryos (Fig. 6). In Stage III and V +ABA embryos, WGA is found in the same tissues locations as reported for mature embryos (Mishkind, Raikhel, Palevitz, Keegstra 1982; Mishkind, Keegstra, Palevitz 1980). It was noted however that Stage V +ABA embryos accumulate extra amounts of WGA in these tissues (data not shown). Hence, +ABA embryos in culture accumulate WGA in the same specific tissue locations as embryos that develop on the plant. Also, when the maturation pathway is bypassed (i.e. -ABA embryos) the resulting seedlings (P) lack WGA in the coleoptiles.

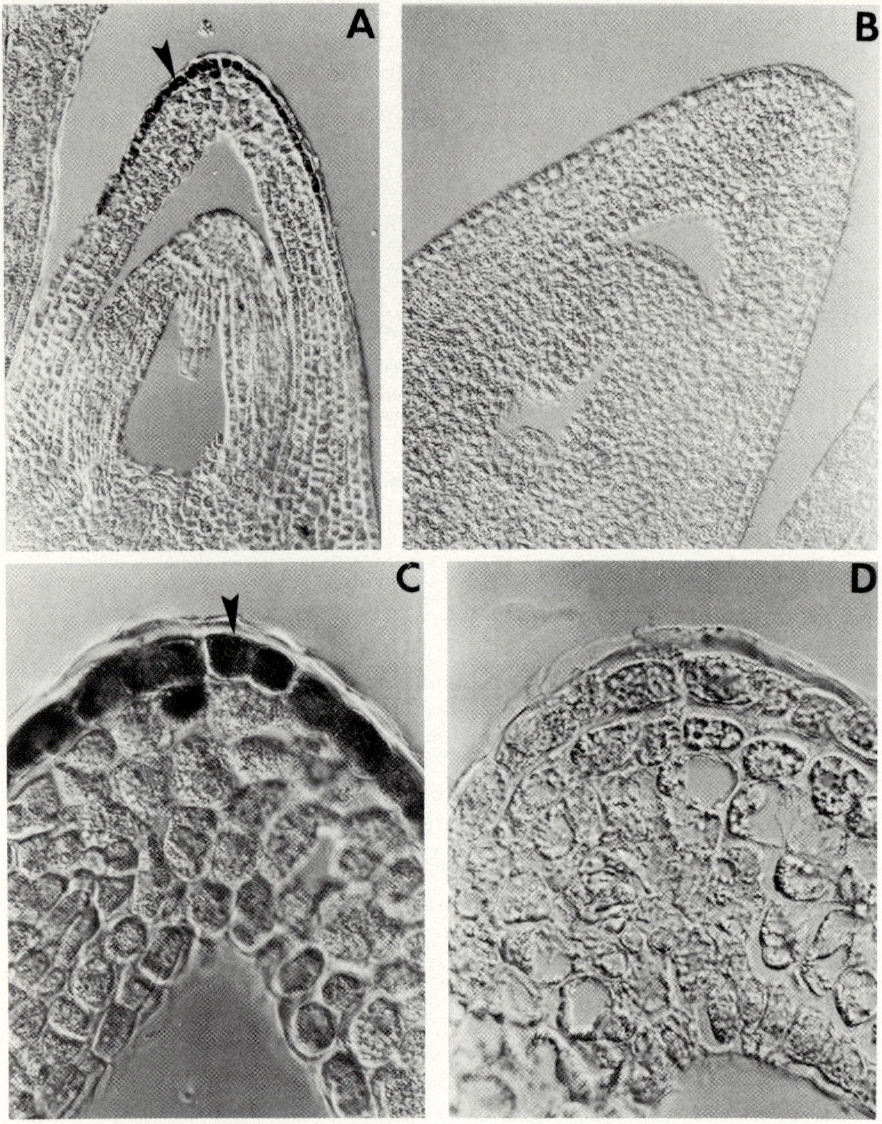

Fig. 6. Sections of shoot tips (A and B) from Stage III Chinese Spring embryos incubated in +ABA (A) and -ABA (B) X190. Sections of coleoptiles (C and D) from the same shoot tips shown in A and B respectively X760. WGA is localized by immunocytochemistry in only the outer layer of the coleoptile in +ABA treated tissue (arrows).

Several possibilities could account for these results. WGA accumulation in the radicle may occur at an earlier developmental stage than accumulation in the coleoptile. Hence, when Stage II or III embryos are removed from the grain, WGA accumulation in the radicle is already programmed. Radicles may be more sensitive to endogenous levels of ABA while exogenous ABA is required for a coleoptile response. Finally, radicles may have an ABA-independent mechanism for WGA accumulation. Also, it is tempting to speculate that there may be two different biochemical forms of WGA (Rice 1976) in the two different tissues (see Fig. 4), and, that WGA localization in the root, unlike that in the coleoptile, is required for the germination process and/or establishment of the seedling. Also, a possible role of WGA in the coleoptile [e.g. to impede fungal growth (Mishkind, Raikhel, Palevitz, Keegstra 1982)] might now be tested by comparing the responses of seedling (M), which has WGA in the coleoptile, to seedling (P) which lacks WGA in the coleoptile.

Ammirato PV (1983). The regulation of somatic embryo development in plant cell cultures: suspension culture techniques and hormone requirements. Biotechnology 1:68.

Chamberlain JP (1979). Fluorographic detection of radioactivity in polyacrylamide gels with the water-soluble fluor, sodium salicylate. Analyt Biochem 98:132.

Crouch ML, Sussex IM (1981). Abscisic acid promotes accumulation of storage proteins in cultured embryos of Brassica napus. Planta 153:64.

Dunwell JM (1981). Dormancy and germination in embryos of Hordeum vulgare L.-Effect of dissection, incubation, temperature and hormone application. Ann Bot 48:203.

Hall TC, Ma Y, Buchbinder BU, Pyne JW, Sun SM, Bliss FA (1978). Messenger RNA for G_1 protein of French bean seeds: Cell-free translation and product characterization. Proc Natl Acad Sci USA 75:3196.

Imamura J, Harada H (1980). Effects of abscisic acid and water stress on the embryo and plantlet formation in anther culture of Nicotiana tabacum cv. Samsun. Z Pflanzenphysiol 100:285.

King RW (1976). Abscisic acid in developing wheat grains and its relationship to grain growth and maturation. Planta 132:43.

Martial JA, Baxter JD, Goodman HM, Seeburg PH (1977). Regulation of growth hormone messenger RNA by thyroid and glucocorticoid hormones. Proc Natl Acad Sci USA 74:1816.

Mishkind M, Keegstra K, Palevitz BA (1980). Distribution of wheat germ agglutinin in young wheat plants. Pl Physiol 66:950.

Mishkind M, Raikhel NV, Palevitz BA, Keegstra K (1982). Immunocytochemical localization of wheat germ agglutinin in wheat. J Cell Biol 92:753.

Peumans WJ, Stinissen HM, Tierens M, Carlier AR (1982). In vitro synthesis of lectins in cell-free extracts from dry wheat and rye embryos. Pl Cell Reports 1:212.

Quatrano RS, Ballo BL, Williamson JD, Hamblin MT, Mansfield M (1983). ABA controlled expression of embryo-specific genes during wheat grain development. In Goldberg RB (ed): "Plant Molecular Biology," New York: Alan R. Liss, in press.

Rice, R (1976). Wheat germ agglutinin. Evidence for a genetic basis of multiple forms. Biochim Biophys Acta 444:175.

Rogers SO, Quatrano RS (1983). Morphological staging of wheat caryopsis development. Amer J Bot 70:308.

Sternberger LA (1979). "Immunocytochemistry". John Wiley and Sons, New York, pp 130-169.

Thomas JO, Kornberg RD (1975). An octamer of histones in chromatin and free in solution. Proc Natl Acad Sci USA 72:2626.

Triplett BA, Quatrano RS (1982). Timing, localization, and control of wheat germ agglutinin synthesis in developing wheat embryos. Devel Biol 90:491.

Umbeck PF, Norstog K (1979). Effects of abscisic acid and ammonium ions on morphogenesis of cultured barley embryos. Bull Torrey Bot Club 106:110.

Chemical Taxonomy, Molecular Biology, and Function of Plant Lectins, pages 131–141

BINDING OF ADENINE AND CYTOKININS TO LIMA BEAN AND OTHER LEGUME LECTINS

David D. Roberts and Irwin J. Goldstein

Department of Biological Chemistry
The University of Michigan
Ann Arbor, Michigan 48109

Lectins are a diverse group of proteins which share the common property of having multiple carbohydrate binding sites. Whereas carbohydrate binding has been the basis for the extensive use of these proteins in studying glycoconjugate structure and function, its relationship to lectin function in plants remains unknown.

Within the Family Leguminoseae evidence has accumulated for conservation of the primary structure of lectins, both in the NH_2-terminal sequences (Foriers et al., 1979) and throughout the primary structure of those lectins for which the complete sequence has been determined (Cunningham et al., 1975, Cunningham et al., 1979, Hopp et al., 1982, Foriers et al., 1981). Regions of homology appear to extend beyond the residues thought to function in carbohydrate and metal ion binding and include residues which in concanavalin A were found to form a hydrophobic pocket which binds nonpolar molecules (Hardman and Ainsworth, 1973).

We have investigated the binding of several nonpolar fluorescent ligands to the blood group A-specific lectin from lima beans (*Phaseolus lunatus*) (Roberts and Goldstein 1982). Binding of 1,8-anilinonaphthalene- sulfonate (ANS) and 2,6-toluidinylnaphthalenesulfonate (TNS) to lima bean lectin (LBL) was shown to be consistent with the existence of two classes of hydrophobic sites. Tetrameric component III (Gould and Scheinberg 1970) has a single high affinity site for TNS and four low affinity TNS sites, one/subunit, which presumably also bind ANS.

TNS has also been reported to bind to concanavalin A with a stoichiometry of one/tetramer (Yang *et al.*, 1974). ANS was found to bind to ricin and *Ricinus communis* agglutinin (Houston 1980). Recently we have investigated the binding of ANS and TNS to a number of legume and nonlegume lectins (Roberts and Goldstein 1983a). ANS sites were found to be conserved in almost all legume lectins. Many legume lectins also bound TNS, with affinity constants generally higher than the affinity of the respective lectins for ANS. In all lectins the hydrophobic ligand binding sites appear to be independent of the carbohydrate binding sites.

The high degree of conservation of hydrophobic sites suggested that physiological analogs of ANS or TNS may be present in plants and that their binding to legume lectins may be related to lectin function. ANS and TNS binding sites on proteins often correspond to binding sites for nonpolar enzyme cofactors or hormones (see Brand and Gohlke 1972 for review). Binding of the plant auxin, indoleacetic acid to hydrophobic sites on concanavalin A has been reported (Edelman and Wang, 1978). However, the affinity constant was low, $K_a = 7 \times 10^2 M^{-1}$, which limits the significance of auxin binding to lectin function. Recently several additional lectins, including those from pea, lentil, *Vicia faba*, and two *Lathyrus* species have been reported to bind indoleacetic acid, although affinity constants were not determined (Chatelain *et al.*, 1982).

A structural similarity exists between the fluorescent probes which bind to LBL (e.g. *N*-phenyl-1-naphthylamine, 1) and a class of adenine derivatives, known as cytokinins (2).

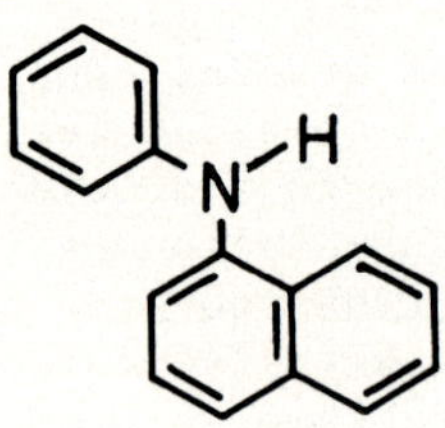

1

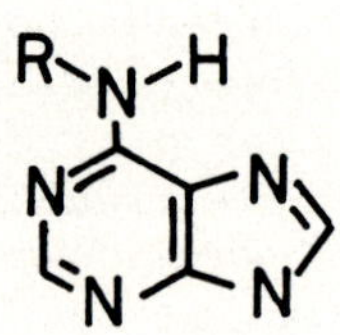

2

The cytokinins generally have unsaturated nonpolar substituents at the _N_-6 position (Leonard 1974) which could be mimiced by the phenyl ring of _N_-phenyl-1-naphthylamine. The cytokinins are phytohormones, and are thought to play an important role in controlling plant growth (Skoog and Armstrong 1970, Letham 1978). In _Phaseolus vulgaris_ and several other dicotyledons, cytokinins have been reported to be important in controlling mobilization of seed reserve proteins during germination (Gepstein and Ilan 1970, 1979, Metivier and Paulilo 1980a, b).

We have used two approaches to examine interaction of LBL with adenine and cytokinins (Roberts and Goldstein 1983b). Equilibrium dialysis experiments using $[^{14}C]$-adenine revealed a single high affinity site on tetrameric LBL component III ($K_d = 1.2 \times 10^{-5}M$). The synthetic cytokinin _N_-6-benzyladenine (BAP) and the naturally occurring cytokinin _trans_-zeatin (6-(4-hydroxy-3-methyl-_trans_- 2-butenylamino)purine) both competed with $[^{14}C]$-adenine for binding to this site. In contrast hypoxanthine, lacking the amino group at the 6- position of the purine ring, failed to bind to the adenine site. These results demonstrated a high degree of specificity for adenine and its _N_-6-substituted derivatives.

The adenine and cytokinin binding site of LBL appears to be identical to the high affinity TNS site identified previously (Roberts and Goldstein 1982). Titration of LBL with TNS in the presence of saturating concentrations of BAP gave a linear Scatchard plot with an affinity constant consistent with binding of TNS only to the low affinity subunit sites (Roberts and Goldstein 1983b). High affinity binding was abolished in the presence of BAP. A proposed model for the lima bean lectin consistent with the observed stoichiometries of cytokinin, ANS, and carbohydrate binding is depicted in Fig. 1.

Based on equilibrium dialysis results, BAP and adenine were not expected to bind to the subunit hydrophobic sites. Remarkably, however, addition of BAP to LBL in the presence of ANS produced a large _increase_ in ANS fluorescence (Roberts and Goldstein 1983b). Titration of an ANS-LBL mixture with BAP demonstrated that the

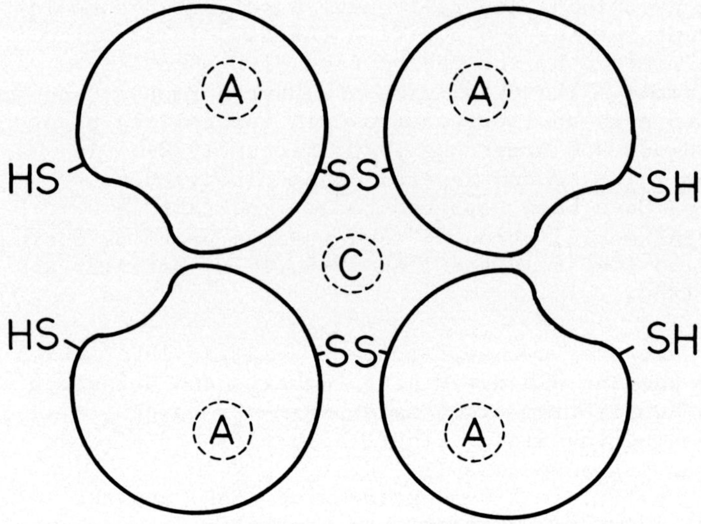

Fig. 1. Model of a lima bean lectin component III tetramer showing ANS (A) and cytokinin (C) binding sites.

enhancement was a saturable phenomenon, where the concentration of BAP producing half- maximal enhancement was comparable to the concentration of BAP needed to half saturate the single high affinity adenine site (Fig. 2). The increase in ANS fluorescence was found to be due to an increase in the affinity of LBL for ANS induced by BAP.

Similar experiments were done using adenine derivatives and cytokinins (Roberts and Goldstein 1983b). Adenine and most cytokinins enhanced ANS binding to LBL. The potency of adenine analogs in this assay correlated well with the affinity of the compounds as measured by equilibrium dialysis. By both methods the affinity of analogs decreased in the order: adenine > benzyladenine > zeatin > hypoxanthine.

To further define the specificity of the adenine binding site we have examined enhancement of ANS binding by a series of substituted adenine analogs (Roberts, Townsend and Goldstein, manuscript in preparation). It is apparent from these studies, that several regions of the

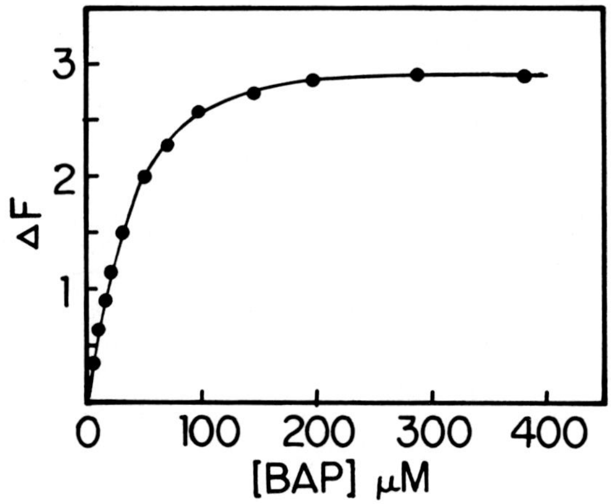

Fig. 2. Effect of benzyladenine on ANS fluorescence enhancement by lima bean lectin. LBL (0.28 mg/ml) containing 1.0 x 10^{-4}M ANS was titrated with benzyladenine at 25°C. A control titration was done in the absence of protein to correct for nonspecific enhancement. Corrections were also made for dilution due to added ligand, excitation at 420 nm, emission at 480 nm.

purine ring are critical to binding, as measured by ability to activate ANS binding. Introduction of methyl substituents at the one and three positions strongly inhibits binding. On the other hand substitution at the two position is well tolerated. Purine or cytokinin nucleosides and purine nucleotides are generally poor ligands. Finally, rearrangement of the 7 and 9 nitrogens to the pyrazolo derivatives alters the apparent affinity. The pyrazolo [4,3-d] pyrimidine analog is more than three times as potent as adenine.

LBL was used as a model system to study the specificity of the hydrophobic binding sites. Since binding of ANS and TNS has been generalized to other legume lectins, we were interested in determining whether binding of adenine derivatives might be a general property of legume lectins and whether the heterotropic

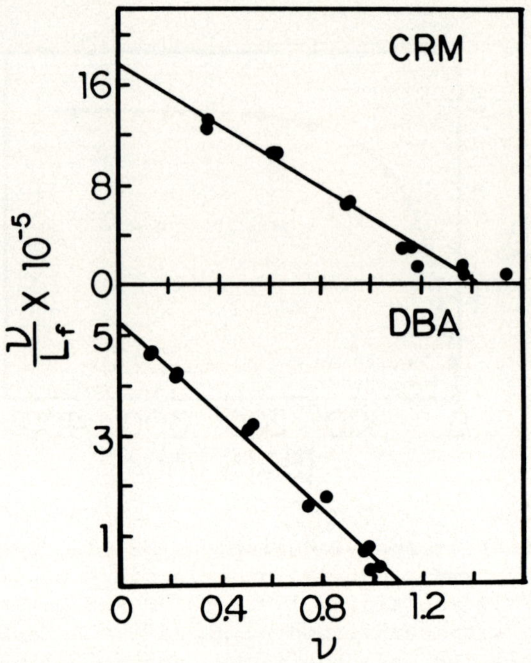

Fig. 3. Equilibrium dialysis with [^{14}C]-adenine. D. biflorus seed lectin and CRM were equilibrated with [^{14}C]-adenine at 25°C, pH 6.8 for 24 hours. Upper panel, 1.25 X 10^{-6}M CRM, determined by radioimmunoassay and assuming M_r = 69,000. Lower panel, 12.5 μM seed lectin, determined by Lowry assay using bovine serum albumin as standard and assuming M_r = 110,000.

interactions described for LBL might also be observed in other lectins. We began by examining the binding of [^{14}C]-adenine to several other legume lectins using equilibrium dialysis. High affinity binding of [^{14}C]-adenine was observed for Dolichos biflorus seed lectin (K_d = 2.0 X 10^{-6}M) and cross reacting material (CRM, Talbot and Etzler 1978) (K_d = 8 X 10^{-7}M), Fig. 3. Soybean agglutinin and P. vulgaris E_4 isolectin also bound adenine with dissociation constants of 1.3 X 10^{-5}M and 8.6 X 10^{-6}M respectively. Griffonia simplicifolia I-A_4 isolectin did not show high affinity adenine

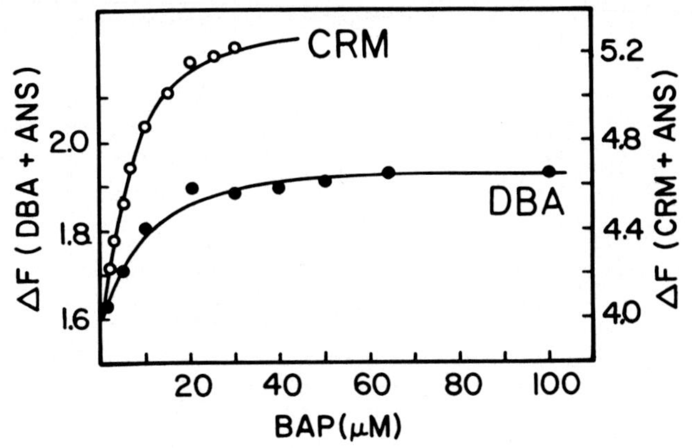

Fig. 4. Effect of benzyladenine on ANS fluorescence enhancement by D. biflorus seed lectin (DBA) and CRM. Conditions were as described in Fig. 2 using 0.25 mg/ml seed lectin and 0.2 mg/ml CRM.

binding. Thus, adenine binding is not an exclusive property of LBL. It is also evident that not all legume lectins have high affinity adenine binding sites. It is notable that the CRM from D. biflorus binds adenine with a higher affinity than the seed lectin. We have previously reported that ANS and TNS bind to both the CRM and seed lectin (Roberts and Goldstein 1983a). The stoichiometry of binding to CRM was based on an estimation of CRM concentration by radioimmunoassay. Whereas the seed lectin bound one adenine/tetramer, the stoichiometry of CRM for adenine could be one or two/dimer.

Heterotropic interactions between ANS and cytokinin binding to other lectins was examined by measuring enhancement of ANS fluorescence following the addition of cytokinins. The D. biflorus seed lectin and CRM behaved in a manner similar to LBL. Addition of BAP to seed lectin or CRM containing 1 x 10^{-4}M ANS produced an increase in ANS fluorescence (Fig. 4). As observed for LBL, the enhancement was saturable. However, half-maximal enhancement occurred at a lower concentration of BAP (5-10 X 10^{-6}M) than was required for LBL.

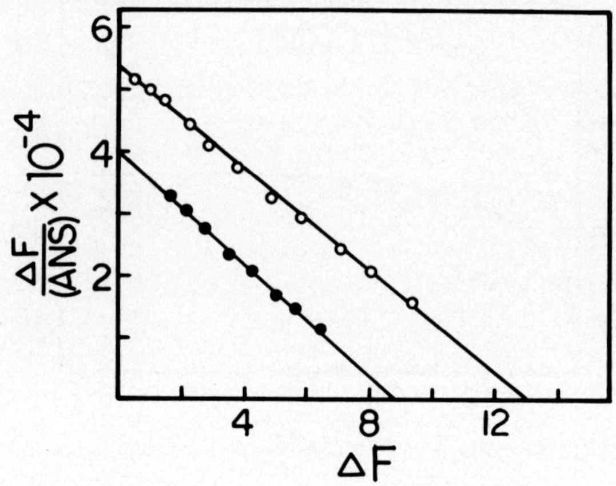

Fig. 5. Fluorescence titration of ANS binding to D. biflorus seed lectin (0.5 mg/ml, ●) and lectin containing 1 X 10^{-4}M benzyladenine (o).

The mechanism of enhancement of ANS fluorescence also differed from that found for LBL. Scatchard plots for titration of D. biflorus seed lectin with ANS in the presence and absence of BAP (Fig. 5) demonstrate that either the number of binding sites or the quantum yield of bound ANS is increased in the presence of BAP, whereas the affinity of ANS binding is not significantly altered.

Interaction between ANS and BAP binding was also examined for P. vulgaris-E_4 and soybean agglutinin. BAP partially inhibited ANS binding to P. vulgaris E_4. However, ANS still bound with positive cooperativity in the presence of BAP, as was reported previously for ANS binding in the absence of cytokinins (Roberts and Goldstein 1983a). No perturbation by BAP of ANS binding to soybean agglutinin was observed. In interpreting the variations in the effect of BAP on ANS binding it is important to note that ANS is not a naturally ocurring ligand. The effect of cytokinin binding on the interactions of possible physiological ANS analogs with lectins could differ from the effects observed with ANS.

However, no ligands other than ANS have yet been found which bind with high affinity to the ANS sites of lectins.

In summary, the presence of hydrophobic binding sites which interact strongly with adenine and its N-6 derivatives has been demonstrated on several legume lectins. This suggests that lectins may not function solely as carbohydrate binders but could also be important as binding proteins for adenine and cytokinins. Since lectins are often present at very high concentrations in seeds, this binding could function to reduce the concentration of free cytokinins in the seed. It should also be considered that the site of function of lectins may be in a tissue other than the seed. In this regard, it is interesting that the D. biflorus CRM, which is isolated from stems and leaves, binds adenine with even greater affinity than the seed lectin. Further work in this area is needed to determine whether lectins or related proteins such as the D. biflorus CRM might function in expression of any of the physiological effects of cytokinins in legumes.

This work was supported by National Institutes of Health Grant GM 29470 and Training Grant 5-T32-GM-07767 and a Rackham Predoctoral Fellowship from the University of Michigan.

REFERENCES

Brand L, Gohlke JR (1972). Fluorescence probes for structure. Ann Rev Biochem 41:843.

Chatelain C, Oustrin J, Rouge P (1982). In vitro interaction between lectins and the plant auxin-indoleacetic acid. Ann Pharm Fr 40:473.

Cunningham BA, Wang JL, Waxdal MJ, Edelman GM (1975). The covalent and three-dimensional structure of concanavalin A. J Biol Chem 250:1503.

Cunningham BA, Hemperly JJ, Hopp TP, Edelman GM (1979). Favin versus concanavalin A: circularly permuted amino acid sequences. Proc Natl Acad Sci USA 76:3218.

Edelman GM, Wang JL (1978). Binding and functional properties of concanavalin A and its derivatives: interactins with indoleacetic acid and other hydrophobic ligands. J Biol Chem 253:3016.

Foriers A, deNeve R, Strosberg AD (1979). Lectin sequences as a tool for chemotaxonomical classification. Physiol Veg 17:597.

Foriers A, Lebrun E, Van Rapenbusch R, deNeve R, Strosberg AD (1981). The structure of the lentil (Lens culinaris) lectin. J Biol Chem 256:5550.

Gepstein S, Ilan I (1970). A promotive action of kinetin on amylase activity in cotyledons of Phaseolus vulgaris. Plant & Cell Physiol 11:819

Gepstein S, Ilan I (1979). Cytokinin-induced amylolytic activity in bean cotyledons: identification of the regulated enzyme. Plant & Cell Physiol 20:1603.

Gould NR, Scheinberg SL (1970). Isolation and partial characterization of two anti-A hemagglutinins from P. lunatus. Arch Biochem Biophys 137:1.

Hardman KD, Ainsworth CF (1973). Binding of nonpolar molecules by crystalline concanavalin A. Biochemistry 12:4442.

Hopp TP, Hemperly JJ, Cunningham BA (1982). Amino acid sequence and variant forms of favin, a lectin from Vicia faba. J Biol Chem 257: 4473.

Houston LL (1980). Differential fluorescence enhancement of 8-anilino-1-naphthalene sulfonic acid by ricin A and B chains. Biochem Biophys Res Commun 92:319.

Leonard NL (1974). Chemistry of the cytokinins. Rec Adv Phytochem 7:21.

Letham DS (1978). Cytokinins. In Letham DS, Goodwin PB, Higgins TJV (eds): "Phytohormones and related compounds: a comprehensive treatise", Amsterdam:Elsevier, p 205.

Metivier J, Paulilo MT (1980a). The utilization of cotyledonary reserves in Phaseolus vulgaris L. cv. carioca. J Exp Bot 31:1257.

Metivier J, Paulilo MT (1980b). The utilization of cotyledonary reserves in Phaseolus vulgaris L. cv. carioca: the effect of 6-benzyladenine and gibberellic acid upon embryonated and detached cotyledons. J Exp Bot 31:1271.

Roberts DD, Goldstein IJ (1982). Hydrophobic binding properties of the lectin from lima beans (Phaseolus lunatus). J. Biol Chem 257:11274.

Roberts DD, Goldstein IJ (1983a). Binding of hydrophobic ligands to plant lectins: titrations with arylaminonaphthalenesulfonates. Arch Biochem Biophys 224:in press.

Roberts DD, Goldstein IJ (1983b). Adenine and cytokinin binding sites of the lectin from lima beans (Phaseolus lunatus). Submitted to J Biol Chem.

Skoog F, Armstrong DJ (1970). Cytokinins. Ann Rev Plant Physiol 21:359.

Talbot CF, Etzler ME (1978). Isolation and characterization of a protein from leaves and stems of Dolichos biflorus that cross reacts with antibodies to the seed lectin. Biochemistry 17:1474.

Yang DCH, Gall WE, Edelman GM (1974). Rotational correlation time of concanavalin A after interaction with a fluorescent probe. J Biol Chem 249:7018.

Chemical Taxonomy, Molecular Biology, and Function of Plant Lectins, pages 143–161

THE POSSIBLE ROLE OF LECTINS IN MEDIATING PLANT CELL-CELL INTERACTIONS

Marilyn A. Anderson, Rosslyn D. Hoggart and
Adrienne E. Clarke
Plant Cell Biology Research Centre,
School of Botany, University of Melbourne
Parkville, Vic. 3052 Australia

Plant cells, like animal cells, can discriminate between self and non-self. This capacity for cell recognition is expressed in a number of ways, for example in the capacity of the female organ, the pistil, to discriminate between compatible and incompatible pollen during fertilization of flowering plants. Somatic cells of flowering plants also have the capacity for cell recognition, as stem grafts can be established between closely related plants, but generally are less successful between distantly related plants. There are also other specific flowering plant-flowering plant relationships such as those between certain plants and their parasites such as mistletoes (Clarke and Knox 1978). Flowering plants also form quite specific symbiotic associations with both fungi (e.g. mycorrhiza) and bacteria (e.g. Rhizobium); other bacteria and fungi are effective pathogens. Some pathogenic interactions are precisely controlled genetically, and the interaction is believed to involve a specific interaction of genes in host and pathogen (Ellingboe 1976).

The cell recognition systems of flowering plants which have been most extensively studied are the economically important legume-Rhizobium interactions and host-pathogen interactions. In both cases, plant lectins are implicated in the recognition between host and microorganism (Schmidt and Bohlool 1981). Much of the work on legume-Rhizobium interactions developed from the idea that as lectins are characteristic components of legumes, they may function to recognize the surface saccharides of the Rhizobium (Bauer 1981). One approach to understanding host-pathogen interactions is similarly based on the idea that pathogens have surface saccharide sequences, which may be recognized by appropriate host lectins (Schmidt and Bohlool 1981). It has been extremely difficult to either prove or disprove these hypotheses, although it is clear that surface saccharide-lectin interactions are involved in some animal cell interactions (Frazer and Glazier 1979; Barondes 1981; Gabriel 1982). To date there is no direct evidence that lectins mediate interactions involving flowering plant cells in contact, although in several cases, lectins are both tissue specific (Kauss 1981) and developmentally regulated (Meimeth et al. 1982) which argues in favour of their having a specific recognition function.

The study of fertilization in flowering plants provides a model, which in contrast to the legume-Rhizobium and host-pathogen interactions, involves a genetically defined recognition event between two cell types of the flowering plant. The question of whether lectins are involved can be approached by defining the protein products of the S-gene which controls the interactions between pistil and pollen tube and then establishing whether this component is or is not a lectin.

BIOLOGY OF FERTILIZATION OF FLOWERING PLANTS

The female organ of flowering plants is the pistil which consists (generally) of the stigma, style and ovary (Fig. 1a). The male gametes (sperm) are carried in pollen which is transported by wind, water or animal vectors to the stigma surface of the pistil. If the mating is compatible, the pollen hydrates and germinates to produce a pollen tube which penetrates the stigma surface and grows intercellularly through the style to the embryo sac where two sperm cells are released, one fertilizes the egg and the other fuses with the primary endosperm nucleus (Fig. 1b). The pistil is able to select between the different types of pollen which it may receive. Intergeneric and intrageneric crosses are usually prevented; intraspecific crosses are often successful, but there are many plant families in which specific self-incompatibility genes (S-genes) operate to prevent in-breeding. In these families there are often no morphological differences between plants in different S-allele groups (for reviews see de Nettancourt 1977; Heslop-Harrison 1978; Lewis 1979; Pandey 1979). Analagous, genetically controlled systems exist for prevention of self-

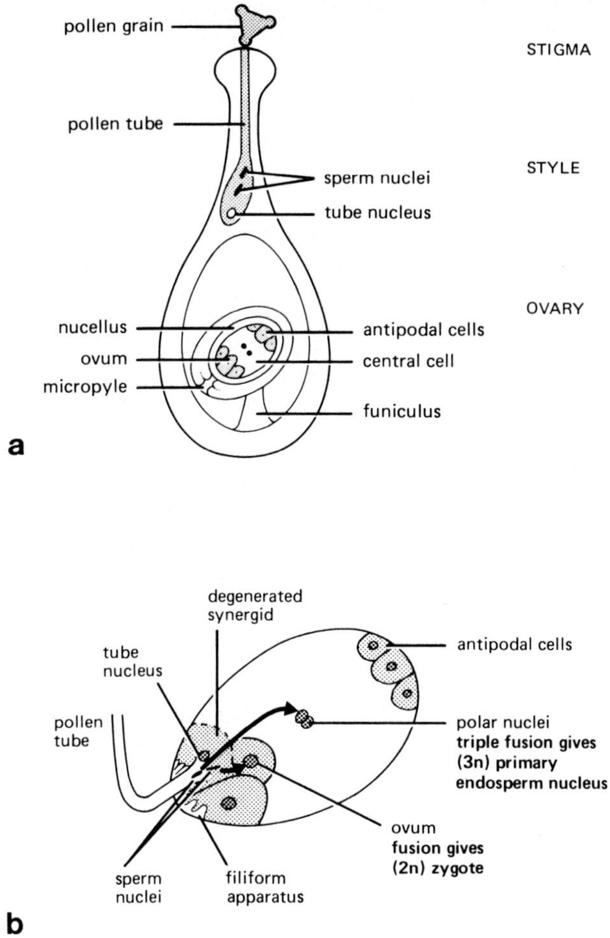

Fig. 1. Schematic representation of fertilization

(a) Compatible pollen germinates at the stigma surface, grows through the style to the ovary.

(b) In the embryo sac, the sperm cells are released from the pollen tube: one fertilizes the egg and the other fuses with the primary endosperm nucleus.

fertilization in some hermaphrodite invertebrates (Schofield et al. 1982). There are two major types of self-incompatibility in plants:

1. Sporophytic Incompatibility

The incompatibility response is controlled by the S-gene, which usually has a single locus with multiple alleles. If an allele of the pollen parent is also present in the female tissues, pollen tube growth is inhibited at the stigma surface (Fig. 2a). The arrest of growth is accompanied by rapid deposition of callose both at the stigma surface adjacent to the point of contact with the tube and also in the tip of the tube. The interaction can be viewed as a cell-cell recognition event in which components of the surfaces of the stigma and either the pollen grain or rudimentary pollen tube, interact to signal either continued or arrested pollen tube growth (Clarke and Gleeson 1981). This type of incompatibility operates for example, in the Cruciferae (cabbage family) and the Compositae (daisy family).

2. Gametophytic Incompatibility

In gametophytic systems, pollen behaviour is controlled by its own genotype and growth of incompatible pollen tubes is arrested not at the stigma surface but within the style. Control is also mediated through a single gene locus (S-locus) with multiple alleles. A pollen tube carrying a single given allele is inhibited if the same allele is present in the style (Fig. 2b). The gametophytic polyallelic system is considered to be the ancestral form of self-incompatibility in angiosperms with the sporophytic system being derived from it (de Nettancourt 1977), so that the products of the S-locus in the two systems may or may not be structurally related. In the gametophytic system both incompatible and compatible pollen grains hydrate and germinate normally. As the pollen tubes grow, callose is deposited at intervals which gives them a ladder-like appearance (Fig. 3). This material is believed to be deposited to cut off the cytoplasm in the growing tip from the spent pollen grain. At some point during growth through the style, usually the upper one-third, the tip of the incompatible pollen tubes becomes abnormal. The normal pollen tube wall has two distinct layers, an outer cellulosic layer, and an inner

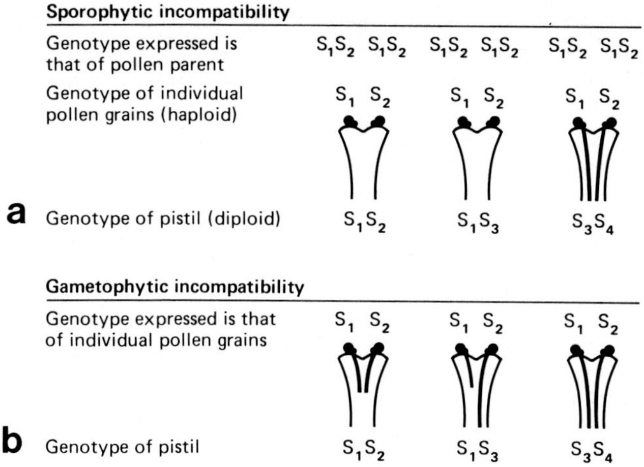

Fig. 2. Behaviour of pollen in the two major self-incompatibility systems.

(a) Sporophytic incompatibility. The pollen parent genotype is S_1S_2. When an allele in the pollen parent is matched with that of the pistil (e.g. S_1S_2 or S_1S_3), pollen germination is arrested at the stigma surface. Where there is no match (S_3S_4) the pollen may germinate and grow through the style to the embryo sac. The central panel only applies if the allele S_1 is dominant to or codominant with S_2 in the pollen and S_1 is dominant to or codominant with S_3 in the style. If S_3 is dominant to S_1 in the style, or S_2 dominant to S_1 in the pollen, the pollen from S_1S_2 parent will be compatible.

(b) Gametophytic incompatibility. The pollen parent genotype is S_1S_2. When an allele in the individual haploid pollen grain is matched with either allele in the diploid stylar tissues, growth of the pollen tube is arrested, usually in the style. For example, both S_1 and S_2 pollen are inhibited in a S_1S_2 style but, unlike the sporophytic system the S_2 pollen will grow successfully through an S_1S_3 style. Where there is no match of alleles (S_1S_2 pollen grains on a S_3S_4 pistil), the pollen tubes of both genotypes grow through the style to the embryo sac.

"callosic" layer (for review see de Nettancourt 1977). The inner callose wall of the pollen tube disappears in incompatible pollen tubes, the outer cellulosic wall enlarges and the tube tip swells and bursts, releasing the sperm prematurely. Often there is a characteristic deposit of callose immediately behind the burst tip (Fig. 3). The cytoplasm at the apex of incompatible tubes is also altered and becomes packed with particles which are released from the burst tip into the intercellular fluid of the transmitting tissue.

Pollen tubes usually either grow through a hollow mucilage-filled style canal or intercellularly through the style transmitting tract. In both instances, the pollen tubes are in contact with and able to interact with extracellular materials in the style. The interaction can be viewed as a cell-cell recognition event in which a component(s) of the extracellular material secreted by the cells of the style is perceived by the pollen tube, the outcome ultimately being either continued or arrested pollen tube growth.

In this chapter, we concentrate our discussion on gametophytic self-incompatibility (GSI) as the genetics are well-defined and relatively simple. It occurs for example in some species of Solanaceae (e.g. petunia, tomato, tobacco), Rosaceae (e.g. cherries) and Leguminoseae (e.g. clover).

AN APPROACH TOWARDS THE UNDERSTANDING OF GAMETOPHYTIC SELF-INCOMPATIBILITY (GSI).

A number of hypotheses have been presented to explain the genetic basis of GSI since self-incompatibility was first compared to the animal immune system (East 1929).

Classical genetics suggests that GSI is controlled by a single dominant gene with multiple alleles. Since the S-gene arose during the evolution of flowering plants to promote outcrossing and as it is expressed in over 70 flowering plant families, it is believed to be highly conserved (for review see de Nettancourt 1977). The prevention of interspecific fertilization is also probably controlled in some way by the S-gene, or a gene closely linked to the S-gene, as the functions of self-incompatibility and interspecific incompatibility cannot be separated (at least in the Solanaceae) by recombination.

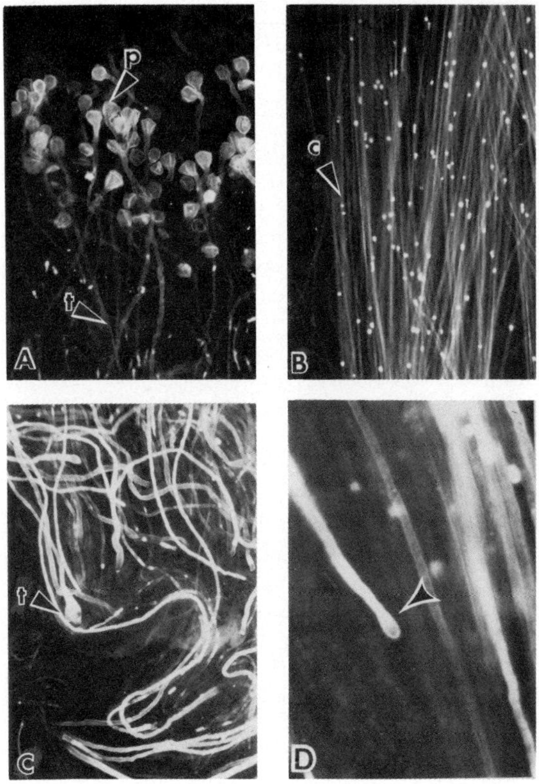

Fig. 3. Appearance of pollen tubes in the style of *Nicotiana alata*

(a) Compatible pollination at the stigma and the upper part of the style. The pollen grains (p) have germinated to produce pollen tubes (t).

(b) Compatible pollination within the upper one-third of the style. The tubes have a ladder-like appearance due to the deposition of callose (c) at regular intervals in the tube walls.

(c) Incompatible pollination within the upper one-third of the style. The tubes (t) grow irregularly and the tips swell and burst.

(d) Detail of an incompatible tube showing a swollen tip (arrow) and the characteristic deposit of callose immediately behind the swollen tip.

The interactions between the extracellular material of the style and pollen tubes during an incompatible pollination can be considered in terms of a specific recognition event and a response to this recognition.

Recognition

Incompatibility occurs if the S-gene products of the pollen tube and the transmitting cells in the style are expressed from the same allele. It is difficult to conceive a mechanism that would lead to arrest of tube growth as a response to interaction of two cell types bearing the same gene product. Several questions need to be answered before this problem can be addressed:

Are the pollen tube and style S-gene products identical? Although the S-gene product from the pollen tube and the stylar cells is generated from the same allele it need not be identical in both cell types. Post-translational modification, specific to each cell type, could generate different proteins for example by different glycosylation or phosphorylation mechanisms or possibly by peptide cleavage. Alternatively, different mRNA's and thus different proteins could be produced from the same allele if different splicing patterns are possible.

Several paradigms are available for this sort of gene expression e.g. Adenovirus type 2 and Simian virus 40 (SV40) are both able to produce more than one mRNA from the same 'gene' or transcription unit due to different splicing patterns. In the case of Adenovirus early region 1A, each of the two major mRNA's in turn, generates two products due to different post-translational processing of the encoded protein (Berk and Sharp 1978; Esche et al. 1980).

What is the subcellular location of the S-gene product arising from both cell types? It is presumed that the stylar cells secrete the S-gene product because the pollen tubes grow through the extracellular matrix making few direct contacts with the walls of transmitting tissue cells. It is not known whether the pollen tube S-gene product is also secreted or if it is retained in the cell wall, the plasma membrane or some other cellular compartment.

This problem of the location of material involved in pollen-style recognition is a particular example of the general problem of how plant cell recognition operates. A major difference between plant and animal cells is that the plasma membrane (and any membrane-bound receptors) is shielded from direct contact with extracellular signals by the cell wall. The development of the cell wall surrounding the plasma membrane was a crucial evolutionary event determining differences in Plant and Animal Kingdoms (Gunning 1983). It is possible that the cell wall which prevented the evolution of cell-cell recognition systems precisely parallel to that of animal cells, may itself play an active role in the recognition process (Ralton and Clarke 1983). A primary receptor may be associated with the plasma membrane or cell wall (Fig. 4) and an extracellular signal may either diffuse through the wall to the membrane receptor, or be altered by wall enzymes or be recognized by wall held receptors. In the case of the pollen tube, there is no information available as to whether the S-gene product is cytoplasmic, membrane-bound, cell wall bound or secreted. If it is not secreted from the pollen tube, it is unlikely to be identical to the style S-gene product as it may have a wall- or membrane-anchoring sequence which is absent from the secreted style material. Information on the subcellular localization of the S-gene product will allow these general ideas to be refined.

Response

The sequence of events which ultimately leads to arrest of pollen tube growth is not defined. The pattern of tube wall growth and callose deposition is altered indicating abnormal wall biosynthesis, and alterations in the cytoplasmic organelles are seen in electron micrographs of incompatible pollen tubes (for discussion see de Nettancourt 1977; Herrero and Dickinson 1981). Other observations such as the effect of Ca^{++} on pollen tubes grown _in vitro_ (Reiss and Herth 1982) indicates another level of control, but no information on the relationship between these effects and the incompatible response are available.

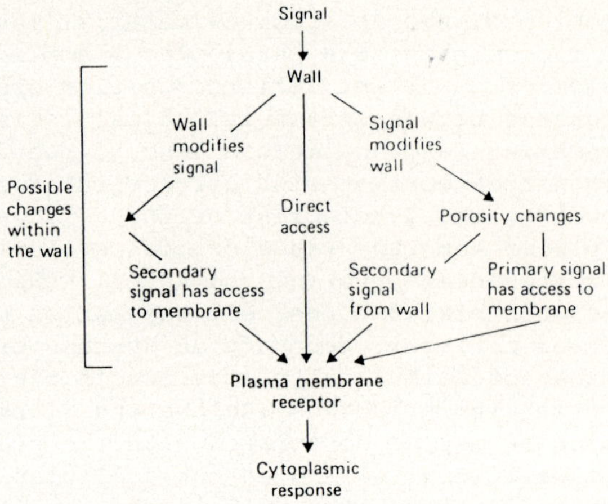

Fig. 4. Possible ways in which the cell wall may be involved in transmission of an extracellular signal to the cytoplasm of a plant cell (from Bacic and Clarke 1983).

The cell wall in flowering plants prevents the direct membrane-membrane contact by which animal cell interactions are mediated. The cell wall may be involved in several ways: it may act merely as a molecular or ionic sieve allowing passive diffusion of molecules of a particular size or charge to complementary receptors. If the extracellular signal is an enzyme, the wall may be modified by hydrolysis and release of wall fragments. Such a modification may induce changes in the wall porosity by modifying the number of junction zones and hence the gel properties of the wall. Alternatively, by changing the wall porosity, a primary signal (enzyme) may gain access to the membrane from which it was initially excluded because of its size. As well as inducing changes in the porosity, the released wall fragment may act as a secondary signal, and interact directly with the plasma membrane receptor. It is also possible that other wall components such as enzymes or lectins may modify the primary signal and in so doing generate a secondary signal which has access to the putative membrane receptor.

A CASE HISTORY: GAMETOPHYTIC SELF-INCOMPATIBILITY IN *NICOTIANA ALATA*

Nicotiana alata, an ornamental tobacco, is an ideal system for examining the molecular basis of pollination. The plants can be maintained in flower throughout the year and the flowers have large accessible pistils. Self-incompatibility is expressed by arrest of 100% of incompatible pollen tubes in a narrow zone within the upper one-third of the style (Fig. 3). Defined S-genotypes are available and homozygotes can be produced by early bud self-pollination of the heterozygotes; this is possible as self-incompatibility is not expressed in the developing bud until the onset of petal colouration.

We have made progress in establishing the basis of recognition during self-pollination in *N. alata*.

1. Electrophoretic Examination of Protein Components of Style extracts

SDS-PAGE of mature style extracts shows identity of bands for all genotypes except in the 30K region where there are individual bands corresponding to genotype. Two-dimensional electrophoresis confirms that genotype-associated bands are present in the 30K region and indicates that they are characteristically of high pI (∿9.0-10.0). Previously, an association of genotype with particular bands in isoelectric focussing gels of style extracts has been noted (Bredemeijer and Blass 1981).

2. Immunological Analyses of Style Extracts Using Rabbit Antisera

Western blots of style extracts indicate that the 30K proteins are major antigens but that there is considerable cross-reactivity of the bands corresponding to different S-genotypes when antisera raised to a particular genotype is used. There are a number of antigenic components in the styles of all genotypes, apart from the S-allele-associated bands in the 30K region. Thus the S-allele-associated proteins in *N. alata* are highly antigenic, like the S-allele associated proteins in other systems examined (for review see Mau et al. 1982).

3. Arabinogalactan-proteins (AGPs)

Arabinogalactans from N. alata (S_1S_3) have been isolated by affinity chromatography using the galactosyl-specific IgA myeloma J539 (Glaudemans 1975). Two families of arabinogalactans are eluted from the affinity column, one with lactose, the other with citric acid; both interact with the Yariv artificial carbohydrate antigen, which is a diagnostic test for AGP's; both are associated with protein which characteristically has a high hydroxyproline content, and both contain Ara and Gal as the major monosaccharides, with the Gal:Ara ratio being higher in the citric acid-eluted fraction. Together, these groups of arabinogalactans account for 82% of the total style carbohydrate (A. Gell and A. Bacic unpublished data).

Arabinogalactan-proteins are quantitatively important components of style extracts (Gleeson and Clarke 1979) and are found in styles of flowers of all major plant families (Hoggart and Clarke 1983). Proteoglycans of this class are not restricted to style tissues but are present in many tissues and secretions of plants from all major families (Clarke et al. 1979; Fincher et al. 1983). They are characterized by a high proportion of carbohydrate in which galactose and arabinose are predominant monosaccharides. A low proportion of protein, typically containing high levels of hydroxyproline is commonly associated (probably covalently) with the arabinogalactan. Analyses of the Lilium and Gladiolus AGP's are consistent with the molecule having a 1,3-linked β-galactan backbone, branched through C(0)6 to β(1→6) galactan side chains some of which carry terminal L-arabinofuranosyl residues (Fig. 5). There are minor differences in the composition of the side chains in the isolated AGP's and some evidence that terminal disaccharides of some arabinogalactan gums have taxonomic significance (Clarke et al. 1979). This implies that the saccharides may be involved in expression of identity of individual plant tissues; it is possible therefore that the side chain composition of style AGP's is also significant. For example, by analogy with blood group substances, the AGP's might have specific terminal glycosyl residues which are specified by glycosyl transferases.

The AGP's were originally regarded as lectins (Jermyn and Yeow 1975; Clarke et al. 1978) on the grounds that they specifically bind the β-glycosyl moiety of the Yariv artificial carbohydrate antigens and that the binding is inhibited by flavanol glycosides. However, although they bind particular glycosyl residues when presented in an appropriate configuration, they do not agglutinate any of the animal cells tested and hence are not formally lectins.

The relationship, if any, between the stylar AGP's and the S-gene products is not known.

4. Lectins in the Style

Because lectins are implicated in recognition in plant host-symbiont, host-pathogen and algal mating systems, as well as in some animal cell recognition systems, it is logical to examine the interacting partners (style, pollen or pollen tube) in self-incompatible pollination for lectins. In three systems haemagglutinins are present in style extracts [*Primula* (Golynskaya et al. 1976), *Ginko* (Pettitt 1977), *Gladiolus* (Gleeson 1980)] but were not detected in *Datura* (Kilpatrick et al. 1979) or *N. alata* styles.

In an intensive search for lectins in *N. alata* styles, no haemagglutinating activity was detected at protein concentrations up to 5.6mg/ml. Nor did the extracts agglutinate Spheron beads to which the monosaccharides Fuc, Ara, Xyl, were coupled. This negative finding is difficult to interpret as it is possible that the assay cocktail lacked essential ions or that an inappropriate sequence of glycosyl residues was present at the red blood cell surface or the surface of the Spheron-monosaccharide beads for detection of a particular lectin. The current definition of lectins (Goldstein et al. 1980) and the methods used for their detection may give a bias away from detecting carbohydrate-binding proteins of biological significance in plants; (as pointed out by Shannon and Hankins 1981) not only are some of the important terminal monosaccharides of plant glycoconjugates such as Rha, Ara, Xyl not present at animal cell surfaces used in agglutination assays, but the multivalent binding required for cell agglutination may not be essential for initiation of a biological effect in plants.

The relationship between lectin and glycosidase activities of *Vigna radiata* (Shannon and Hankins 1981) and *Vica faba* (Dey et al. 1982) seed proteins is of particular interest with regard to the potential recognition function of lectins in plants. One form of the *V. faba* α-galactosidase has lectin activity with glucose/mannose specificity and the carbohydrate-binding sites for catalytic and lectin functions are apparently at separate loci on the same protein (Dey et al. 1982). It is possible that two such binding sites on a single molecule could function separately *in vivo* in mediating cell recognition.

5. Cell Wall Glycoproteins

Another group of Ara, Gal containing glycoconjugates which are implicated in other plant recognition interactions are the hydroxyproline-rich cell-wall glycoproteins. Lectins of the Solanaceae resemble the cell wall glycoprotein in their structure, but in contrast to the cell wall glycoprotein, are water-soluble. The close structural similarity suggests a biological relationship between the two groups which is not clear. Although these glycoproteins resemble the AGP's in having a high Hyp content as well as Ara and Gal as the major monosaccharides, they are structurally quite distinct in the carbohydrate-protein linkage and in the arrangement of monosaccharides within the side chains (Fincher et al. 1983) (Fig. 5). The finding that molecules of this class are determinants of mating in the unicellular algae *Chlamydomonas* (Cooper et al. 1983) indicates that they can have a recognition function and their potential involvement in control of fertilization in higher plants should be considered. In *N. tabacum*, a third Hyp-rich, Ara, Gal containing glycoprotein, which is neither the cell wall glycoprotein nor the $(GlcNAc)_2$ specific lectin, is implicated in species-specific agglutination of pathogenic but not non-pathogenic bacteria (Mellon and Helgeson 1982).

Neither the cell wall glycoprotein nor a Hyp-rich lectin has been described in *N. alata* pollen tubes or styles. However we anticipate that the cell wall glycoprotein, being a universal component of cell walls, would be present and that as *N. alata* is a member of the Solanaceae, a lectin may be present in some tissues. However, as pointed out in the previous section, we have been unable to detect haemagglutinating activity in *N. alata* style extracts.

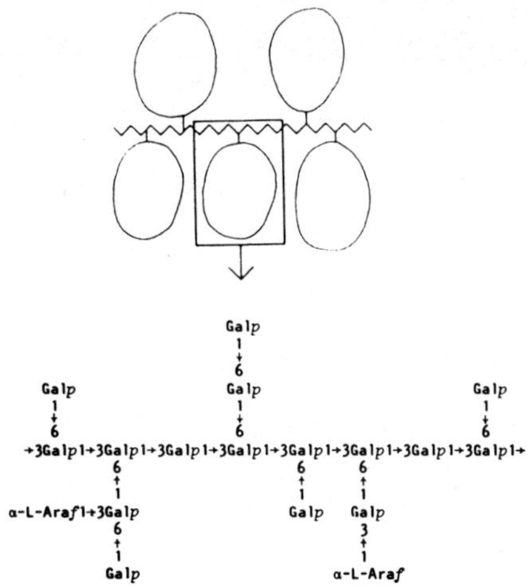

(a) Arabinogalactan-protein

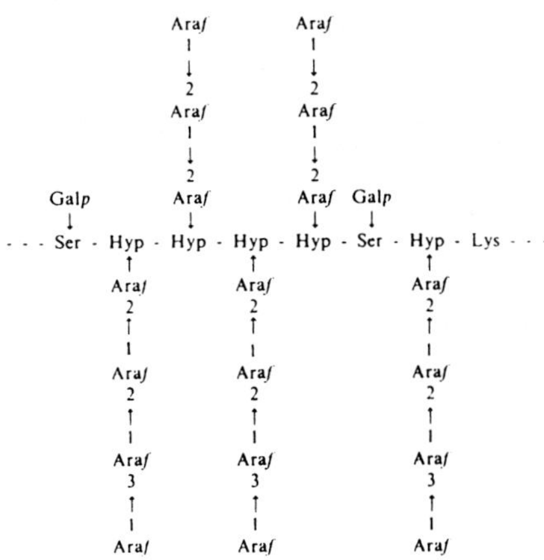

(b) Cell wall glycoprotein

Fig. 5. Possible structures of an arabinogalactan-protein and a cell wall glycoprotein as reviewed by Clarke et al. (1979) and Fincher et al. (1983).

In conclusion, the question of whether lectins may be involved in the recognition between style and pollen tube cannot be answered on the basis of available data. The direct way to resolve the question is to isolate the S-gene product and establish whether or not it is a lectin.

CONCLUSIONS

1. Self-incompatibility (SI) is a plant cell recognition system which is well defined, biologically, genetically and microscopically.
2. The relatively simple genetics of the system (incompatibility controlled by a single dominant S-gene with multiple alleles) makes it amenable to examination by immunological and recombinant DNA techniques. Some progress in identifying putative S-gene products in a number of systems has been made.
3. The molecular basis of recognition in SI will be established via isolation and characterisation of the allelic products of the S-gene.
4. SI is an excellent model for establishing principles of recognition between plant cells which are likely to apply to more complex interactions.
5. The possible role of lectins can be defined by following the S-gene action and establishing at which point lectins are involved.

ACKNOWLEDGEMENTS

It is our pleasure to acknowledge the assistance of Dr K.K. Pandey, DSIR, Grasslands Division, Palmerston North, New Zealand for generously providing us with his _N. alata_ seed stocks. We are most grateful to our colleagues, especially Dr Elizabeth Williams for her direct contribution to many of the ideas presented here, for providing us with the micrographs of Figure 3 and for her continuing support and collaboration during the experimental work.

Dr Philip Harris kindly provided us with critical comments on the manuscript at short notice, and we are grateful for his help.

REFERENCES

Bacic A, Clarke AE (1983). Plants and Immunity. In Moore R (ed): "Vegetative Compatibility Responses in Plants". Proc Am Bot Soc (in press).

Barondes SH (1981). Lectins: Their multiple endogenous cellular functions. Ann Rev Biochem 50:297.

Bauer WD (1981). Infection of legumes by Rhizobia. Ann Rev Plant Physiol 32:407.

Berk AJ, Sharp PA (1978). Spliced early mRNAs of simian virus 40. Proc Natl Acad Sci USA 75:1274.

Bredemeijer GMM, Blass J (1981). S-specific proteins in styles of self-compatible Nicotiana alata. Theor Appl Genet 59:185.

Clarke AE, Gleeson PA (1981). The molecular basis of recognition and response in the pollen-stigma interaction. In "Recent Advances in Phytochemistry" Vol. 15, The phytochemistry of cell recognition and cell surface interactions. New York & London: Plenum Press, p 161.

Clarke AE, Gleeson PA, Jermyn MA, Knox RB (1978). Characterization and localization of β-lectins in lower and higher plants. Aust J Plant Physiol 5:707.

Clarke AE, Anderson R, Stone BA (1979). Form and function of arabinogalactans and arabinogalactan-proteins. Phytochemistry 18:521.

Clarke AE, Knox RB (1978). Cell recognition in plants. Quart Rev Biol 53:3.

Cooper JB, Adair WS, Mecham RP, Goodenough UW (1983). The Chlamydomonas agglutinin is a hydroxyproline rich glycoprotein. Proc Natl Acad Sci (submitted).

de Nettancourt D (1977). "Incompatibility in Angiosperms" Berlin, Heidelberg, New York: Springer-Verlag.

Dey PM, Naik S, Pridham JB (1982). The lectin nature of α-galactosidases from Vicia faba seeds. Febs Lett 150(1): 87.

East EM (1929). Self-sterility. Bibliogr Genetica 5:331.

Ellingboe AH (1976) Genetics of host-parasite interactions. In Heitefuss R, Williams PH (eds) "Encyclopaedia of Plant Physiology" Vol. 4. Berlin: Springer-Verlag p 761.

Esche H, Mathews MB, Lewis JB (1980). Proteins and messenger RNAs of the transforming region of wild-type and mutant adenoviruses. J Mol Biol 142:399.

Fincher GB, Stone BA, Clarke AE (1983). Arabinogalactan-proteins - structure, biosynthesis and function. Ann Rev Plant Physiol 34:47.

Frazier W, Glaser L (1979). Surface components and cell

recognition. Ann Rev Biochem 48:491.
Gabriel O (1982). Carbohydrates and receptor recognition. In Kohn LD (ed) "Hormone Receptors", John Wiley & Sons p 137.
Glaudemans CPJ (1975). The interaction of homogeneous, murine myeloma immunoglobulins with polysaccharide antigens. Adv Carbohydr Chem Biochem 31:313.
Gleeson PA (1980). Characterization of carbohydrate polymers associated with pollination in Gladiolus and Secale. PhD Thesis, University of Melbourne.
Gleeson PA, Clarke AE (1979). Structural studies on the arabinogalactan-protein from the style canal of Gladiolus gandavensis. Biochem J 181:607.
Goldstein IJ, Hughes RC, Mosigny M, Osawa T, Sharon N (1980). What should be called a lectin? Nature 285:66.
Golynskaya EL, Bashirova NV, Tomchuk NN (1976). Phytohaemagglutinins from the pistil of primula as possible proteins of generative incompatibility. Fiziol Rast 23(1):88.
Gunning BES (1983). The special features of plant cells. In Watson J, Raff M, Alberts B, Bray D, Roberts K (eds) "Molecular Biology of the Cell", New York: Garland Inc.
Herrero M, Dickinson HG (1981). Pollen tube development in Petunia hybrida following compatible and incompatible intraspecific matings. J Cell Sci 47:365.
Heslop-Harrison J (1978). Genetics and physiology of angiosperms incompatibility systems. Proc Roy Soc Lond B 202:73.
Hoggart RM, Clarke AE (1983). Presence of arabinogalactans in crude stylar extracts of Angiosperms. Phytochemistry (Submitted).
Jermyn MA, Yeow YM (1975). A class of lectins present in the tissues of seed plants. Aust J Plant Physiol 2:501.
Kauss H (1981). Lectins and their physiological role in slime moulds and in higher plants. In Tanner W, Loewus FA (eds) "Encyclopeadia of Plant Physiology" New Series Vol. 13B, Plant Carbohydrates II. Berlin: Springer-Verlag.
Kilpatrick DC, Yeoman MM, Gould AR (1979). Tissue and subcellular distribution of the lectin from Datura stramonium (Thorn Apple). Biochem J 184:215
Lewis D (1979). Sexual Incompatibility in Plants. The Institute of Biology's Studies in Biology No. 110. Edward Arnold.
Mau S-L, Raff JW, Clarke AE (1982). Isolation and partial characterisation of components of Prunus avium L. styles, including an antigenic glycoprotein associated with self - incompatibility genotype. Planta 156:505.
Meimeth T, Tran Thanh Van K, Marcotte J-L, Trinh T-H, Clarke AE (1982). Distribution of lectins in tissues, derived callus and roots of Psophocarpus tetragonolobus (winged bean). Plant Physiol 70:579.

Mellon JE, Helgeson JP (1982). Interaction of a hydroxyproline rich glycoprotein from tobacco callus with potential pathogens. Plant Physiol 70:401.

Pandey KK (1979). Overcoming incompatibility and promoting genetic recombination in flowering plants. NZ J Bot 17:645.

Pettitt J (1977). Detection in primitive gymnosperms of proteins and glycoproteins of possible significance in reproduction. Nature 266:530.

Ralton JE, Clarke AE (1982). Comparison of plant and animal recognition systems. In Greenberg AH (ed) "Cell Receptors and Cell Communication in Invertebrates." Marcel Dekker.

Reiss H-D, Herth W (1982). Disoriented growth of pollen tubes of Lilium longiflorum Thunb. induced by prolonged treatment with the calcium-chelating antibiotic, chlorotetracycline. Planta 156:218.

Schmidt EL, Bohlool BB (1981). The role of lectins in symbiotic plant-microbe interactions. In Tanner W, Loewus FA (eds) "Encyclopaedia of Plant Physiology" New Series Vol. 13B Plant Carbohydrates II. Berlin: Springer-Verlag.

Scofield VL, Schlumpberger JM, West LA, Weissman IL (1982). Protochordate allorecognition is controlled by a MITC-like gene system. Nature 295:400.

Shannon LM, Hankins CN (1981). Enzymic properties of phytohaemagglutinins. Recent Adv Phytochem 15:93.

Chemical Taxonomy, Molecular Biology, and Function of Plant Lectins, pages 163–176

THE CELL BIOLOGY OF WHEAT GERM AGGLUTININ AND RELATED LECTINS

M. L. Mishkind[1], N. V. Raikhel[1], B. A. Palevitz[1], and K. Keegstra[2]

[1]Botany Department, University of Georgia, Athens, GA 30602 and [2]Botany Department, University of Wisconsin, Madison, WI 53706

The report of Mirelman et al. (1975) on the fungastatic properties of wheat germ agglutinin (WGA) stimulated wide interest in the function of this lectin in wheat plants. The authors found that purified lectin binds to hyphal tips of *Trichoderma viride* and *Fusarium solani* and inhibits sodium acetate incorporation, conidiation and hyphal elongation. Chitotriose, a potent hapten of WGA, blocks these effects. They suggested that in nature perhaps "WGA protects wheat against chitin-containing phytopathogens during seed imbibition, germination and early seedling growth".

At the outset of our studies, the quantity and location of WGA during the wheat life cycle was not known. We believed that such information would aid in the evaluation of suggested functions of WGA as well as indicate new areas of investigation. We thus developed methods to quantitate and locate WGA in wheat embryos and plants.

Assays based on the hemagglutination and immunological properties of the lectin were employed. Reports of legume proteins that are not hemagglutinins but possess antigenic determinants similar to those of seed lectins (Mialonier et al. 1973, Talbot and Etzler 1978) necessitated the use of these complimentary procedures. In wheat, we have found that proteins antigenically similar to WGA are present in adult plants and that these proteins possess sugar-binding activity. Thus, passage of a crude extract of wheat tissue over a column of immobilized N-acetylglucosamine (GlcNAc)

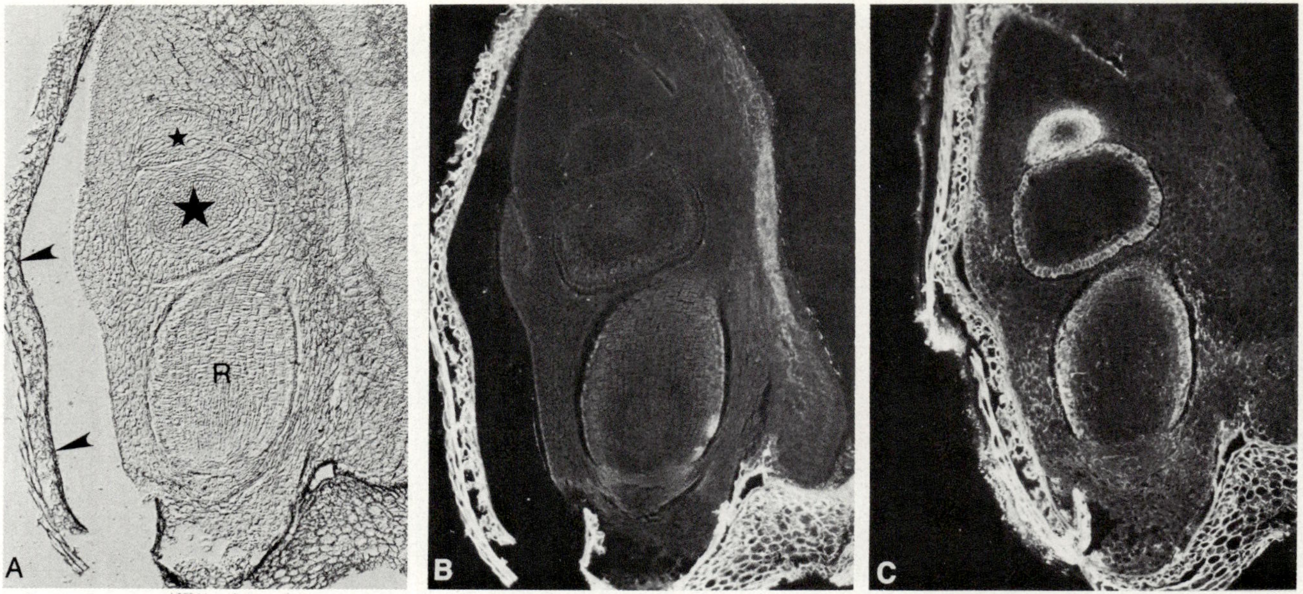

Fig. 1. Immunofluorescence localization of WGA in wheat embryo sections. The radicle (R) appears longitudinally and the adventitious roots (stars) are in cross-section. Sections were photographed with differential interference contrast optics (DIC, panel A) and with fluorescence optics (B, C) for visualization of rhodamine fluorescence. The grain coat (arrowheads), somewhat detached in the section treated with control IgG's (panel B) is strongly autofluorescent (B, C). In the section treated with antiserum to WGA (C), specific fluorescence is visible at the surface of the radical and one of the adventitious roots (large star). The plane of section passes through a second adventitious root near its tip revealing widespread reaction product. From Mishkind et al. (1982). X47

oligomers removes activity detected by radioimmunoassay (RIA, Mishkind et al. 1980) or enzyme-linked immunosorbant assay (ELISA, N. V. Raikhel et al. in preparation). Furthermore, these extracts display hemagglutination titers commensurate with their antigenically-determined levels.

WHEAT GERM AGGLUTININ IN WHEAT EMBRYOS

Minetti et al. (1976) showed by hemagglutination assay that WGA is confined to the embryo portion of the grain. Similar partitioning also has been reported by our laboratory (Mishkind et al. 1980) and others (Stinissen et al. 1981, Triplett and Quatrano 1982). We found by RIA that five hexaploid wheat varieties contain 1.7 to 3.0 μg of WGA per grain. Similar levels, also determined by RIA, were reported by Triplett and Quatrano (1982), whereas Peumans et al. (1982a) found somewhat lower amounts (0.2 μg) by hemagglutination assay.

WGA is located only in particular organs and cell layers of the embryo. Light microscope level immunocytochemistry (Mishkind et al. 1982) showed that in embryos WGA is limited to regions that establish contact with the soil during germination. The surface layers of various organs including the radicle, the first adventitious roots, the coleoptile, and scutellum contain lectin (Fig. 1). Although found throughout the coleorhiza and epiblast, WGA is at highest levels within cells at the surface of these organs. Thus, a small population of cells in the embryo possesses most of the lectin present in the grain. Lectin concentration at these sites therefore must be quite high despite the fact that overall, WGA represents 1% or less of the total grain protein.

DISTRIBUTION OF WGA DURING GERMINATION AND IN ADULT PLANTS

Upon imbibition of the grain, some lectin is released into the surrounding medium. Since only small amounts are recovered (Mishkind et al. 1980, Stinissen et al. 1981), the release may result from a nonspecific loss of protein from the expanding embryos and may not be of physiological significance.

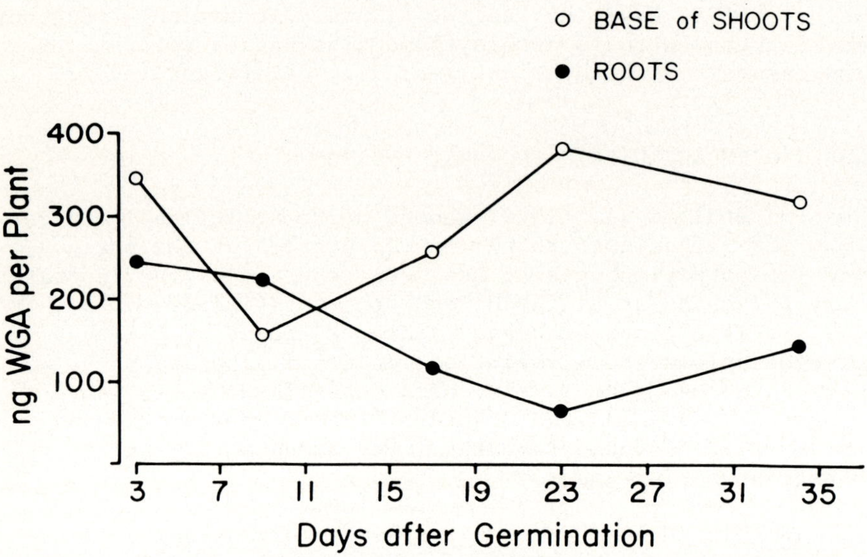

Fig. 2. Level of WGA in wheat plants during the first month of growth. At indicated intervals plants were separated into roots and shoots. The entire root system and the basal 5 cm of shoot were extracted for assay. For day 3 plants only, the entire shoot was assayed. Lectin levels were determined by RIA as described by Mishkind et al. (1980). From data presented in Mishkind et al. (1980).

During the first month of growth the amount of WGA in the plant remains high relative to levels found in the ungerminated grain. We dissected 3 to 34 day old plants into three fractions - leaves, basal region of the shoot containing the meristems, and roots. WGA levels in each fraction were determined by RIA. Although lectin levels fluctuated in the shoot base and root regions, they remained between 60 and 400 ng per plant throughout the time course (Fig. 2). Recently, a sensitive double antibody ELISA has confirmed these findings (N. V. Raikhel et al., in preparation). Although the RIA detected no antigenically cross-reactive material in leaves, the ELISA indicates that low levels are present in this organ. Preliminary results (Haigler et al. in preparation) indicate

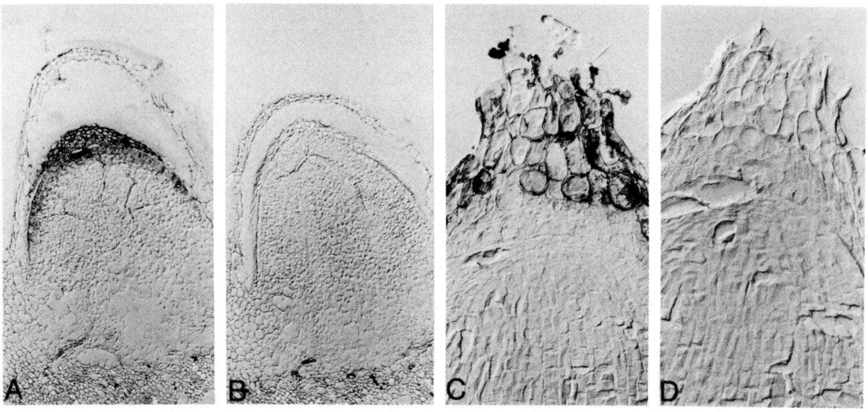

Fig. 3. Localization of a WGA-like lectin in the adventitious roots of two month-old wheat plants. Sections were incubated with anti-WGA (A, C) or control IgG's (B, D) and photographed with DIC optics. Antibody-binding to sections was detected by the peroxidase-antiperoxidase procedure. Note that in roots that have not emerged from the shoot (A) as well as in emerged roots (C) reaction product is restricted to the root tip and cap. From Mishkind et al. (1982). A, B, X33; C, D, X48.

that the level of lectin in leaves is controlled hormonally. Since abscisic acid induces WGA synthesis in developing and germinating embryos (Triplett and Quatrano 1982), we questioned whether this hormone stimulates lectin accumulation in adult plants. Leaves from plants grown for 15 days in nutrient medium supplemented with 10^{-5}M abscisic acid were found to contain five to ten-fold more antigenically cross-reactive material than untreated controls. Although the roots of adult plants contain functional lectin (see below), antigenically-reactive material from leaves has not yet been similarly characterized.

As in the embryo, WGA in adult plants is strictly sequestered into particular organs and cells. Initially this was shown by RIA of root systems separated into young, recently-emerged adventitious roots and older roots (Mishkind et al. 1980). On fresh weight and protein bases, the young roots were determined to be enriched in lectin more than ten-fold over the older roots. By immunocytochemistry we found that compartmentalization is the rule at the cellular as well as the organ level. Before and after emergence from the shoot, adventitious roots contain lectin mainly in the cap and in cells at the root surface (Fig. 3). RIA demonstrated that lectin at the root surface can be eluted by washing intact roots in buffer (Mishkind et al. 1980). Exhaustive washing solubilizes 90 ng of WGA per plant, or about half of the lectin present in the roots. Since addition of GlcNAc oligomers to the wash buffers enhances WGA elution by three-fold, it is likely that WGA interacts with the root surface through its sugar binding sites. The association thus may be through root surface oligosaccharides and/or lectin self-association at its secondary sugar binding site (Wright 1980).

The WGA-like material from adult plants is similar to the embryo lectin in that it agglutinates red blood cells in a GlcNAc-inhibitable manner and it cross-reacts with antibodies raised to the embryo lectin. The adult lectin, as purified by sugar- or antibody-mediated affinity chromatography, co-migrates with the 18 kilodalton embryo lectin subunits on SDS polyacrylamide gels. Characterization of the adult lectin is in progress in our laboratory.

INTRACELLULAR LOCATION OF EMBRYO AND ADULT LECTINS

Immunocytochemistry at the ultrastructural level with ferritin (Mishkind et al. 1982) and colloidal gold (Raikhel et al. submitted) markers demonstrates that WGA and related lectins are restricted to the lumen of protein bodies in the embryo and to lumen of vacuoles in adventitious roots. At both sites, immunospecific reaction product is limited to the peripheral regions close to the membranes of these organelles (Figs. 4 and 5). Since protein bodies and vacuoles are considered to be homologous (Lott 1980), lectin synthesis and packaging steps may be similar in embryos and adults.

Since wheat embryos have not been extensively characterized at the ultrastructural level, we tentatively assign the term protein bodies to the WGA-containing organelles. We note, however, that these structures are

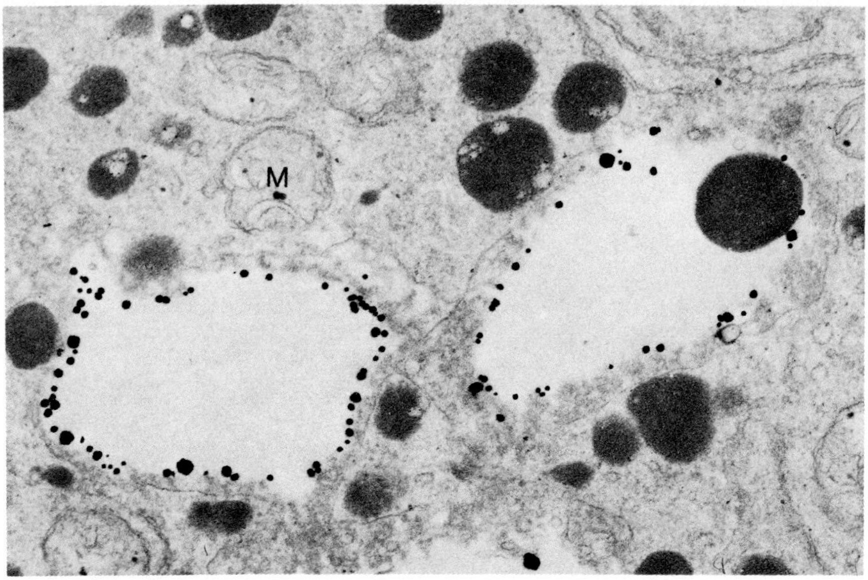

Fig. 4. Intracellular localization of WGA in a mature, imbibed wheat embryo. The section was incubated with anti-WGA and colloidal gold-labeled secondary antibodies prior to preparation for electron microscopy. Note the gold particles are evident only in the lumen of protein bodies. Naturally-occurring electron-dense particles are visible in mitochondria (M). From Raikhel et al., (submitted). X37,000

different from endosperm protein bodies in that they stain only weakly for protein (M. L. Mishkind, unpublished observation).

Pulse-labeling studies of Stinissen et al. (1982a) indicate that the proteolytic processing and transport of WGA precursors across cellular membranes occurs post-translationally. These data are in accord with our inability, despite much effort, to locate WGA within the various compartments of the cellular endomembrane system (N. V. Raikhel, unpublished observations). Hence, it appears that WGA may be synthesized on soluble polyribosomes and enter preformed protein bodies. Structures at the periphery of these organelles (receptors of WGA?)

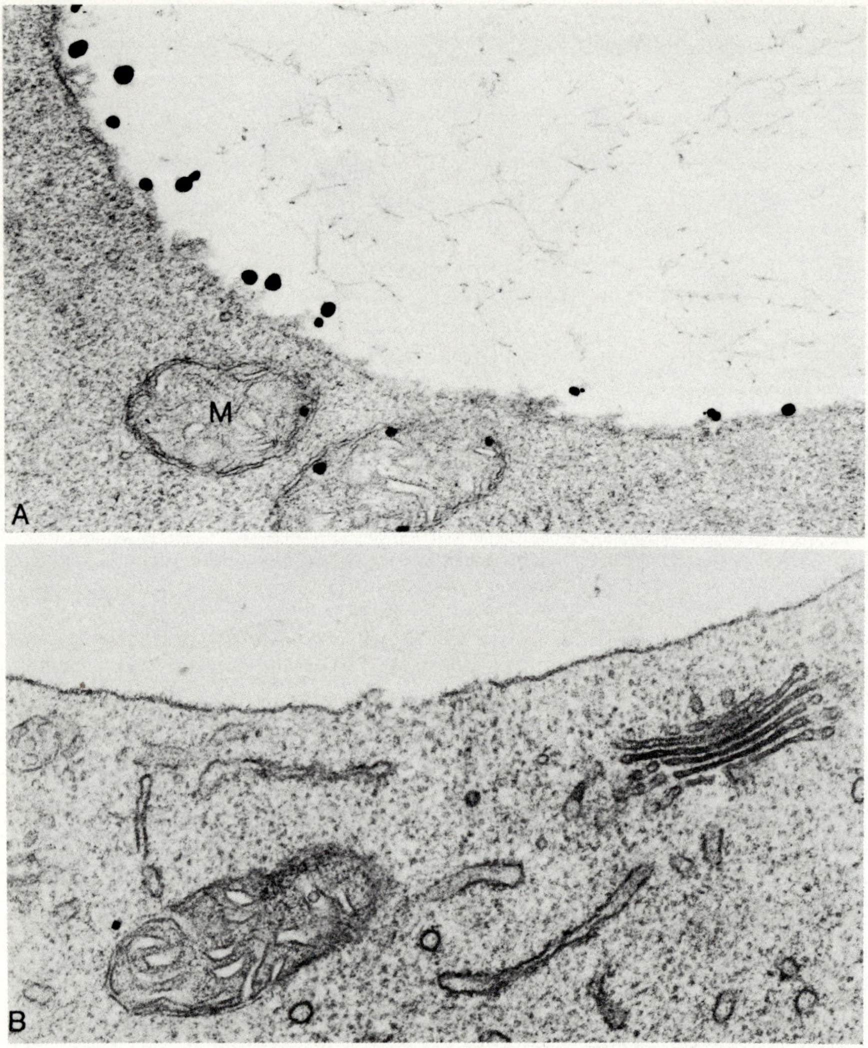

Fig. 5. Intracellular localization of a WGA-like lectin in adventitious roots of adult wheat plants. Sections were treated with anti-WGA (A) or control IgG's (B) and colloidal gold-labeled secondary antibodies as described by Raikhel et al. (submitted). Note that gold particles are evident at the periphery of a vacuole (A) in the anti-WGA-treated tissue and that reaction product is absent in the control. Mitochondria contain naturally-occuring electron-dense deposits (M). A, X57,000; B, X60,000

may be responsible for the striking localization pattern observed.

WGA-LIKE LECTINS IN EMBRYOS OF VARIOUS GRASS SPECIES

Barley and rye embryo lectins are indistinguishable from WGA by immunological and sugar-binding criteria (Mishkind 1981, Peumans et al. 1982b). Their similarity in amino acid composition (Peumans et al. 1982b) further demonstrates the high degree of evolutionary conservatism of these proteins. We performed immunocytological analysis to determine whether the localization patterns of these proteins in embryos is similarly conserved (Mishkind et al. 1983). Both barley and rye embryos, like wheat embryos, were found to contain lectin at the periphery of the radicle and in the coleorhiza. In contrast, localization patterns in the coleoptiles varies between species (Fig. 6). Whereas lectin in the wheat coleoptile is limited to the outer surface of this organ, rye coleoptiles contain lectin in both the inner and external cell layers. Unlike wheat and rye, barley lacks reaction product in the coleoptile.

A lectin similar to WGA in amino acid composition and sugar-binding properties was purified from rice bran by Tsuda (1979). Since rice and wheat are only distant relatives in the grass family, we sought to determine whether antibodies to WGA cross-react with the rice lectin. Although no precipitin reaction occurs between a rice bran extract and antiserum to WGA on Ouchterlony double diffusion plates, antiserum to WGA does bind immunospecifically to sections of rice embryos. Generally, the localization patterns in rice and wheat are similar in that reaction product is concentrated in peripheral regions, ie. the coleoptile, coleorhiza, root cap and periphery of the radicle. In rice however, lectin is more extensively distributed (Fig. 7). All cells of the coleoptile, a number of internal cell layers of the radicle, as well as occasional leaf cells contain lectin. These differences in localization pattern and the inability of antibody to WGA to precipitate rice lectin upon immunodiffusion may reflect the evolutionary distance between rice and wheat.

Lectins similar to WGA were not detected in all the cereals assayed. Oat embryos do not contain extractable hemagglutinating activity nor an antigen cross-reactive

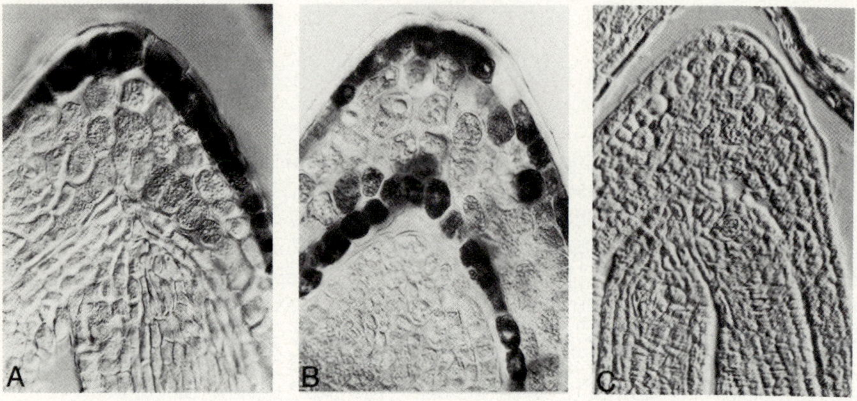

Fig. 6. Localization of WGA-like lectins in wheat, rye, and barley coleoptiles. Sections were treated with anti-WGA and photographed with DIC optics. Antibody-binding to sections was detected with the peroxidase-antiperoxidase procedure. Note that a WGA-like protein is present in the outer cell layer of the wheat (A) coleoptile, in both inner and outer cell layers of the rye (B) coleoptile, but is absent from this organ in barley (C). X200

with antibodies to WGA (Mishkind et al. 1983). Similarly, pearl millet embryos lack reaction product when incubated with anti-WGA. Thus, WGA-like lectins appear to be restricted to only certain tribes in the grass family. The presence of such a lectin in rice, a member of a tribe thought to be primitive by morphological criteria (Stebbins, 1982), suggests that WGA-like lectins may be a primitive trait only selectively retained during the evolution of the Gramineae.

COMPARISON OF WGA AND THE LEGUME LECTINS

Although they are hemagglutinins and are found in seeds, WGA and related lectins clearly represent a different class of proteins from the well-characterized legume seed lectins. Present at microgram quantities per grain, WGA is in low abundance relative to the milligram levels of lectin present in most legume seeds. Their low levels as well as their absence from the endosperm argue

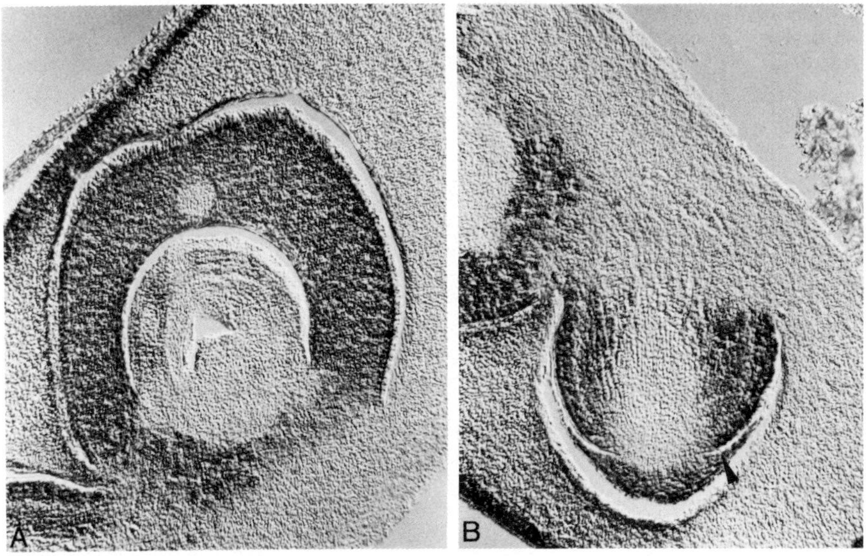

Fig. 7. Localization of a WGA-like lectin in a rice embryo. Sections through the coleoptile and leaves (A) and radicle (B) were incubated with antiserum to WGA and photographed with DIC optics. Antibody-binding to sections was detected by the peroxidase-antiperoxidase procedure. Note that immunospecific reaction product is found throughout the coleoptile and in the root cap (arrow) and peripheral cells layer of the radicle. X420

against a role for the WGA-like lectins as storage proteins or as antiherbivore agents as has been suggested for the legume seed lectins (Janzen et al. 1976).

Unlike the legume lectins, WGA levels remain high (relative to the level in ungerminated grain) during the first month of growth. While legume lectins dissappear with kinetics similar to those of storage proteins, WGA levels are, in comparison, quite stable (Fig. 8).

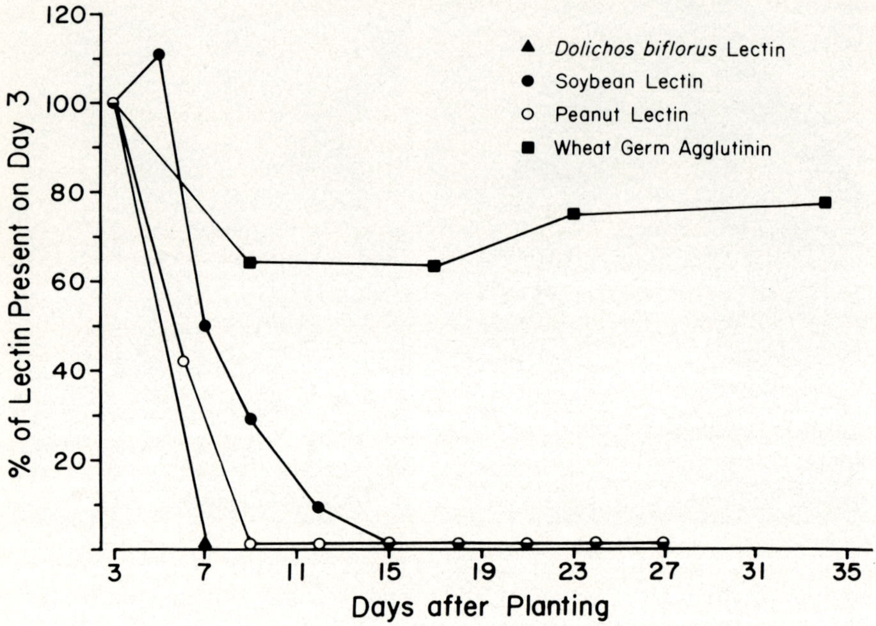

Fig. 8. Relative lectin levels in various plant species during the first month of growth. Levels are expressed as the percentage of lectin present on the third day after imbibition. Values are from Talbot and Etzler (1978), Peuppke et al. (1978), Peuppke (1979), and Mishkind et al. (1980) for *Dolichos bifloris*, soybean, peanut and wheat lectins, respectively.

CONCLUSIONS

During their evolutionary divergence the WGA-like lectins retained not only the ability to cross-link sugars, but specifically to bind the oligomers of GlcNAc present in chitin. In contrast, the legume seed lectins display some variety in their sugar binding specificities. Thus, when discussing potential functions of the WGA-like lectins, one must come to terms with their remarkable affinity for chitin. Since WGA is fungastatic the suggestion of Mirelman et al. (1975) that WGA functions in the defense against fungal pathogens remains compelling. Our data strengthen their contention in that WGA appears limited in both the embryo and adult plant to structures that come in close contact with the outside environment. By slowing fungal growth in regions of possible entry into the plant such as occur during germination or when adventitious roots penetrate the epidermis, time might be gained for the activation of more permanent physical or biochemical barriers. The high concentration of WGA at these sties may be sufficient to achieve these ends.

These suggestions are of course tentative and demand direct experimental test. For this reason other hypothetical functions, such as a role in seed dormancy as proposed by Stinissen et al. (1982b), also must be considered.

ACKNOWLEDGMENTS

We thank Dr. C. Haigler for unpublished data, Dr. G. W. Schmidt for stimulating and critical discussions, and S. Kroehnke for printing the micrographs. Supported by NSF grants PCM81-04470 to BAP and PCM77-25399 to KK.

REFERENCES

Janzen DH, Juster HB, Liener IE (1976). Insecticidal action of the phytohemmagglutinin in black beans on a Bruchid beetle. Science 192:795.

Lott JNA (1980). Protein bodies. In Tolbert NE (ed): "The Biochemistry of Plants," New York: Academic Press, p 589.

Mialonier G, Privat J-P, Monsigny M, Kahlem G, Durand R, Welther ML (1973). Isolement, propriétés physico-chimiques et localization in vivo d'une phytohemagglutine (lectine) de Phaseolus vulgaris L. (var. rouge). Physiol Vég 11:519.

Minetti M, Aducci P, Teichner A (1976). A new agglutinating activity from wheat flour inhibited by tryptophan. Biochim Biophys Acta 437:517.

Mirelman DE, Galun E, Sharon N, Lotan R (1975). Inhibition of fungal growth by wheat germ agglutinin. Nature 256:414.

Mishkind M, Keegstra K, Palevitz BA (1980). Distribution of wheat germ agglutination in young wheat plants. Plant Physiol 66:950.

Mishkind ML (1981). Dissertation. State University of New York at Stony Brook.

Mishkind M, Raikhel NV, Palevitz BA, Keegstra K (1982). Immunocytochemical localization of wheat germ agglutinin in wheat. J Cell Biol 92:753.

Mishkind M, Palevitz BA, Raikhel NV, Keegstra K (1983). Localization of wheat germ agglutinin-like lectins in various species of the Gramineae. Science, in press.

Peumans WJ, Stinissen HM, Carlier AR (1982). Lectin synthesis in developing and germination wheat and rye embryos. Planta 156:41.

Peumans WJ, Stinissen HM, Carlier AR (1982b). Isolation and partial characterization of wheat-germ-agglutinin-like lectins from rye (*Secale cereale*) and barley (*Hordeum vulgare*) embryos. Biochem J 203:239.

Pueppke SG (1979). Distribution of lectins in the Jumbo Virginia and Spanish varieties of the peanut, *Arachis hypogaea* L. Plant Physiol 64:575.

Peuppke SG, Bauer WD, Keegstra K, Ferguson AL (1978). Role of lectins in plant-micro-organism interactions. II. Distribution of soybean lectin in tissues of *Glycine max* (L.) Merr. Plant Physiol 61:779.

Stebbins GL (1982). Major trends of evolution in the Poaceae and their possible significance. In Estes JR, Tyrl RJ, Brunken JN (eds): "Grasses and Grasslands", Norman: University of Oklahoma Press, p 3.

Stinissen HM, Peumans WJ, Carlier AR (1981). Some biological aspects of the wheat-embryo lectin. Arch Int Physiol Biochim 89:B132.

Stinissen HM, Peumans WJ, Carlier AR (1982a). In vivo synthesis and processing of cereal lectins. Plant Mol Biol 1:277.

Stinissen HM, Peumans WJ, Vrijdaghs AJ, Carlier AR (1982b). Accumulation and synthesis of lectins in embryos of developing cereals. Arch Int Physiol Biochim 90:B69.

Talbot CF, Etzler M (1978). Development and distribution of *Dolichos bifloris* lectin as measured by radioimmunoassay. Plant Physiol 61:847.

Triplett BA, Quatrano RS (1982). Timing, localization, and control of wheat germ agglutinin synthesis in developing wheat embryos. Devel Biol 91:491.

Tsuda M (1979). Purification and characterization of a lectin from rice bran. J Biochem 86:1451.

Wright CS (1980). Crystallographic elucidation of the saccharide binding mode in wheat germ agglutinin and its biological significance. J Mol Biol 141:267.

Chemical Taxonomy, Molecular Biology, and Function of Plant Lectins, pages 177–185

HYDROXYPROLINE-RICH GLYCOPROTEINS AND LECTINS IN THE DEFENSE OF PLANTS

Marie-Thérèse Esquerré-Tugayé

Centre de Physiologie Végétale - LA 241 CNRS
Université Paul Sabatier
118, route de Narbonne 31062 Toulouse, France

Since the discovery of the accumulation of hydroxy-proline-rich glycoproteins (HRGP) in the cell wall of diseased plants (Esquerré-Tugayé 1973), and of the presence of hydroxyproline (Hyp) in some lectins (Allen, Neuberger 1973), there have been a number of studies on the possible involvement of Hyp-containing molecules in plant-microorganism interactions.The high insolubility of HRGP once in the cell wall is a limitation to directly assessing its role in plants. However, analysis of different structural markers, together with recent progress in the isolation and identification of a putative precursor of the wall HRGP enables HRGP to be studied in different situations.

This paper reports on the characterization of HRGP's in plants, and summarizes the present knowledge on cell wall HRGP and Hyp-containing lectins in different plant-pathogen systems. A comparison between HRGP and a lectin within a given system is also presented.

I - CHARACTERIZATION OF HRGP's IN PLANTS

A variety of Hyp-containing glycoproteins have been identified in green plants. They include cell wall HRGP's of lower (*Chlamydomonas*) and higher plants, arabinogalactan proteins,(AGP,usually found outside the cell wall,for instance in the growth medium of cells in culture) and lectins of the Solanaceae family (Allen , Neuberger 1973 ; Kilpatrick 1980 ; Lamport 1965 ; Roberts 1974 ; Stuart, Varner 1980 ; Van Holst, Klis 1981). They all have in common a high level of

Hyp and serine, and of either glycine or alanine. In these proteins, Hyp is substituted by oligosaccharides composed of arabinose, galactose and glucose. The cell wall HRGP is characterized by repeated peptides containing the unique sequence -Ser-Hyp_4- in which Ser may be glycosylated by 1 α-D galactopyranosyl residue and Hyp by 1 to 4 residues of L-arabinofuranose (Akiyama et al. 1980 ; Esquerré-Tugayé, Lamport 1979 ; Lamport 1973). The potato lectin also contains the same Hyp-oligoarabinosides (Ashford et al. 1982). Analysis of Hyp-Ara and Ser-Gal gives a good picture of the wall HRGP in a given situation.

A quick estimation of HRGP can be achieved by measuring the amount of Hyp within plant cell walls. It is also possible to measure HRGP biosynthesis after incubation of plant tissues and cells with ^{14}C-proline since Pro is the direct precursor of Hyp in proteins (Esquerré-Tugayé, Mazau 1981 ; Sadava et al. 1973). This technique was notably used for the isolation of a HRGP precursor (Brysk, Chrispeels 1972).

Possible precursors of the wall HRGP have now been isolated from carrot roots, tobacco and melon callus, and potato tubers (Leach et al. 1982 ; Mellon, Helgeson 1982 ; Mazau, Esquerré-Tugayé unpublished data ; Stuart, Varner 1980). They contain exceptionnally high levels of Hyp (40 to 50 %), of tyrosine, and of the basic amino acids histidine and lysine, which confers on them a basic pI. They are also glycosylated : 30 to 60 % sugars for a Mol.Wt. of 90 000 to 120 000. The melon HRGP precursor has the same glycosilation pattern as the wall HRGP. The carrot HRGP seems to have the polyproline II structure (Varner 1982).

Whether the different HRGP's mentioned above are different members of the same family of Hyp-containing molecules, or represent fragments of the same molecule is not known. It has been suggested that the potato Hyp-lectin has at least 2 or 3 domains, one of them being rich in Hyp but devoid of lectin activity, while the rest of the molecule lacks Hyp and bears the lectin activity (Neuberger 1982). A similar hypothesis was proposed by Lamport (1977) who also postulated the existence of a Hyp-rich, and a Hyp-lacking domain in the wall HRGP,the latter being possibly related to the extracellular AGP. If a labile link exists between the two domains, one would expect to split it under the harsh conditions (Ethanol-1.25 M HCl) most often used for the extraction of the wall HRGP precursor. Molecular and immunochemical approaches will probably help to clarify the relationships between Hyp-containing molecules, in the future.

The above reported structural features have been retained to study HRGP and Hyp-lectins in different plant-microorganism systems.

II - CELL WALL HRGP AND HYP-CONTAINING LECTINS IN PLANT-MICROORGANISM INTERACTIONS

1 - CELL WALL HRGP in PLANT-MICROORGANISM INTERACTIONS

In 1973, we reported that a large increase of Hyp occurs in the cell wall of melon seedlings inoculated with *Colletotrichum lagenarium*. Because of no race-cultivar specificity in this system, the significance of Hyp accumulation was not clear. In order to elucidate this point the following approaches were used : a, HRGP in different plant-microorganism systems ; b, HRGP biosynthesis and induced resistance or susceptibility ; c, HRGP elicitation in plants.

a, HRGP in different plant-microorganism systems. Several plants inoculated with either fungi, bacteria or viruses were analyzed. All the dicotyledonous plants studied presented an increase of Hyp and Hyp-oligoarabinosides when infected. The magnitude of the response apparently varies according to the plant family and to the type of disease. Fungi of the genus *Colletotrichum*, for instance, induce a high response in their hosts (Esquerré-Tugayé et al. 1980) ; likewise, different plants of the Cucurbit family accumulate large amounts of HRGP upon infection by different fungi. HRGP is also enhanced more rapidly in cultivars showing race specific resistance, and is associated with the non-nodulation of alfalfa roots in the presence of high NO_3^- concentrations in the medium. The results obtained with these different systems are in favor of HRGP accumulation being a general response, a response possibly involved in the plant's general defence. This hypothesis was studied in the melon-*C. lagenarium* system.

b, HRGP biosynthesis and induced resistance or susceptibility. A close correlation exists in the melon-*C. lagenarium* system between the level of cell wall HRGP and the defence of plants against the *C. lagenarium*.

This was shown by artificially modifying the level of HRGP, either with ethylene which promotes the accumulation of HRGP in the cell wall, or with Hydroxyproline which inhibits the biosynthesis of HRGP (Cleland 1967 ; Esquerré-Tugayé et al. 1979 ; Ridge, Osborne 1970). Under these conditions, it appears that inducing high levels of HRGP prior to inoculation also induces the resistance of plants, while lowering the level of HRGP upon infection induces a higher susceptibility to *C. lagenarium*.

These experiments clearly indicate that, as for other defence mechanisms in plants, early enhancement of HRGP is a prerequisite to its efficiency in the protection of plants. Fungal elicitors can be used for this purpose.

c, HRGP elicitation in plants. Ethylene is abundantly produced during the infection of melons and is involved in the stimulation of HRGP biosynthesis (Toppan et al. 1982). Elicitation of ethylene in plants can be performed by treating them with glycoconjugates isolated from the fungal cell surface (Esquerré-Tugayé et al. 1978 ; Paradies et al. 1979 ; Toppan, Esquerré-Tugayé 1983). The elicitors of ethylene isolated from *C. lagenarium* also elicit the synthesis of HRGP (Roby et al. 1983), and the resistance of melon plants against subsequent infection by *C. lagenarium*. Elicitors of different origin are also able to trigger HRGP biosynthesis.

The results obtained with the three different approaches reported demonstrate that increased HRGP deposition in the cell wall during infection or treatment with elicitors is associated to a defence mechanism of plants against pathogens. The nature of this mechanism is as yet unknown Research on Hyp-containing lectins and agglutinins in plant-pathogen interactions may help to clarify the function of HRGP.

2 - HYP-CONTAINING LECTINS in PLANT-PATHOGEN INTERACTIONS

Most studies in this area have been devoted to the interaction between the potato lectin and pathogens of potato. In 1977 Sequeira and Graham reported that a purified lectin preparation agglutinated avirulent, but not virulent, strains of *Pseudomonas solanacearum* and precipitated the cell wall lipopolysaccharide (LPS) of the bacteria. All precipitations were hapten specific : as expected for the potato lectin (Allen, Neuberger 1973), they were inhibited by oligomers of N-acetyl-D-glucosamine. They were also prevented by the addition of the extracellular polysaccharide (EPS) which is formed by virulent, but is absent in avirulent, cells. Due to the presence of EPS, the LPS of virulent cells was not accessible to the lectin ; as a consequence, virulent bacteria did not attach and were free to multiply and invade the host. It was proposed that, in the potato-*Pseudomonas solanacearum* system, the potato lectin mediates species-specific recognition and defence.

No species-specificity, however, was observed in the interaction between potato lectin and *Phytophtora infestans* another pathogen of potato. Cytospores of compatible and incompatible races of the fungus were agglutinated to the same degree by the lectins isolated from a resistant (R_1 gene) and susceptible (r gene) cultivar (Furuichi et al. 1980). The binding phenomenon was inhibited by oligomers of N-acetyl-D-glucosamine, as was the lysis of *P. infestans* zoospores by the lectin, reported by Garas and Kuć 1981. In the latter case, lectins from 4 potato cultivars with different R-genes for resistance to *P. infestans* lysed zoospores from 5 races of the fungus, regardless of the resistance of the cultivar.

The potato lectin also inhibits the growth of *Botrytis cinerea* (Callow 1977).

Recently a reappraisal of the lectin-*Pseudomonas solanacearum* interaction was published by Leach and Sequeira (1982) who found that, upon purification of the potato extract, agglutination of avirulent bacteria was associated with a HRGP-agglutinin and not to the Hyp-potato lectin hemagglutinin. Agglutination was presumably mediated by charge-charge interaction between the polycation-HRGP and the negatively charged LPS, and did not correspond to the specificity of the lectin. These results are similar to those obtained by Mellon and Helgeson (1982) with a tobacco HRGP which agglutinates, under certain ionic strength conditions, avirulent *P. solanacearum* as well as compatible or incompatible races of *Phytophtora infestans* var. *nicotianeae*.

The agglutinating activity of HRGP is, as yet, unclear, but apparently differs from that of Hyp-lectins such as the potato lectin. It is apparent from the reported data, that either HRGP-agglutinin or Hyp-lectin should be as pure as possible for a proper interpretation of the results. This, however, does not preclude that different kinds of agglutinating activities may coexist in the same plant.

III - HRGP AND LECTINS IN A PLANT-PARASITE SYSTEM

We employed the melon-*C. lagenarium* interaction for a comparative study of wall HRGP and of lectins. Preliminary results indicate that the melon HRGP mentioned earlier in this paper is devoid of hemagglutinating activity. On the other hand, a protein with hemagglutinating properties but devoid of Hyp, is present in the sap of melon stems. The lectin, which was purified by affinity chromatography has a specificity for N-acetyl-D-glucosamine oligomers as already reported for lectins isolated from different species of the Cucurbitaceae (Allen 1979),and for the elicitors of *C. lagenarium* ; in addition, it also agglutinates conidia of different fungi, including *C. lagenarium*. This affinity could explain that the lectin decreases sharply upon infection of melon seedlings. Research is now in progress to assess its possible function, and the function of the wall HRGP, in the defence of plants against *C. lagenarium*.

IV - CONCLUSION

More studies are needed to clearly understand the role of HRGP and of Hyp-lectins in plant-microorganism interactions.

Because of the affinity of Hyp-lectins for oligomers of N-acetyl-D-glucosamine, several authors have suggested that they could be involved in specific recognition of microorganisms containing glucosamine in their cell wall. The data presently available,however,indicates that their interaction with pathogens is not species specific. Since most (if not all) studies have been performed with soluble lectins, one needs to know the amount and location of these molecules before any conclusion can be drawn. This approach should also be made with other lectins, the interaction of which with pathogens is still less documented (Eltzler 1981).

The accumulation of HRGP in the cell wall during infection, now demonstrated in various systems, is associated with the general defense of plants. It is likely that the wall HRGP interacts with the anionic materials present on the cell surface of pathogens in a non-species-specific manner.

REFERENCES

Akiyama Y, Mori M, Kato K (1980). ^{13}C-NMR Analysis of Hydroxyproline Arabinosides from *Nicotiana tabacum*. Agric Biol Chem 44:2487.

Allen AK (1979). A lectin from the exudate of the fruit of the vegetable Marrow (*Cucurbita pepo*) that has a specificity for β-1,4 linked N-Acetylglucosamine oligosaccharides. Biochem J 183:133.

Allen AK, Neuberger A (1973). The purification and properties of the lectin from potato tubers, a hydroxyproline-containing lectin. Biochem J 135:307.

Ashford D, Desai NN, Allen AK, Neuberger A, O'Neill M, Selvendran RR (1982). Structural studies of the carbohydrate moieties of lectins from potato *(Solanum tuberosum)* tubers and thorn-apple *(Datura stramonium)* seeds. Biochem J 201:199.

Brysk MM, Chrispeels MJ (1972). Isolation and partial characterization of a hydroxyproline-rich cell wall glycoprotein and its cytoplasmic precursor. Biochim Biophys Acta 257:421.

Callow JA (1977). Recognition, resistance and the role of plant lectins in host-parasite interactions. Adv Bot Res 4:1.

Esquerré-Tugayé MT (1973). Influence d'une maladie parasitaire sur la teneur en hydroxyproline des parois cellulaires d'épicotyles et de pétioles de plantes de Melon. CR Acad Sc Paris Serie D 276:525.

Esquerré-Tugayé MT, Toppan A (1978). Triggering by fungal components of a defense mechanism involving cell wall glycoproteins in plants. 3rd Int Congress Plant Pathology Abstr:229.

Esquerré-Tugayé MT, Lamport DTA (1979). Cell surfaces in plant-microorganism interactions.I. A structural investigation of cell wall hydroxyproline-rich glycoproteins which accumulate in fungus-infected plants. Plant Physiol 64:314.

Esquerré-Tugayé MT, Lafitte C, Mazau D, Toppan A, Touzé A (1979). Cell surfaces in plant-microorganism interactions. II. Evidence for the accumulation of hydroxyproline-rich glycoproteins in the cell wall of diseased plants as a defence mechanism. Plant Physiol 64:320.

Esquerré-Tugayé MT, Mazau D, Toppan A, Roby D (1980). Elicitation, via l'éthylène, de la synthèse de glycoprotéines pariétales associées à la défense des plantes. Ann Phytopathol 12:403

Esquerré-Tugayé MT, Mazau D (1981). Les glycoprotéines à hydroxyproline de la paroi végétale. Physiol Vég 19:415.

Etzler ME (1981). Are lectins involved in plant-fungus interactions ? Phytopath 71:744.

Furuichi N, Tomiyama K, Doke N (1980). The role of potato lectin in the binding of germ tubes of *Phytophtora infestans* to potato cell membrane. Physiol Plant Pathol 16:249.

Garas NA, Kuc J (1981). Potato lectin lyses zoospores of *Phytophtora infestans* and precipitates elicitors of terpenoid accumulation produced by the fungus. Physiol Plant Pathol 18:227.

Kilpatrick DC (1980). Purification and some properties of a lectin from the fruit juice of the Tomato *(Lycopersicon esculentum)*. Biochem J 185:269.

Lamport DTA (1965). The protein component of primary cell walls. Adv Bot Res 2:151.

Lamport DTA (1973). The glycopeptide linkage of extensin : O-D-galactosyl serine and O-L-arabinosyl hydroxyproline. In Loewus F ed, Biogenesis of Plant cell wall polysaccharides. Acad Press New York : 149.

Lamport DTA (1977). Structure, biosynthesis and significance of cell wall glycoproteins. Recent Adv Phytochem 11:79.

Leach JE, Cantrell MA, Sequeira L (1982). Hydroxyproline-rich bacterial agglutinin from potato : Extraction, purification and characterization. Plant Physiol 70:1353.

Mellon JE, Helgeson JP (1982). Interaction of a hydroxyproline-rich glycoprotein from tobacco callus with potential pathogens. Plant Physiol 70:401.

Neuberger A (1982). Hydroxyproline-containing lectins in plants : their structure and activity. South Afr J Sci 78:348.

Paradies I, Humme B, Hope HH, Heitefuss R, Eltsner EF (1979). Induction of ethylene formation in Bean *(Phaseolus vulgaris)* hypocotyl segments by preparations isolated from germn tube cell walls of *Uromyces phaseoli*. Planta 146:193.

Roberts K (1974). Crystalline glycoprotein cell walls of Algae : Their structure, composition and assembly. Phil Trans R Soc Lond B 268:129.
Roby D, Toppan A, Esquerré-Tugayé MT (1982). Cell surfaces in plant-microorganism interactions. V. Elicitors of fungal and plant origin trigger the synthesis of ethylene and of cell wall hydroxyproline-rich glycoproteins in plants. Plant Physiol (submitted).
Sadava D, Walker F, Chrispeels MJ (1973). Hydroxyproline-rich cell wall protein (extensin) : biosynthesis and accumulation in growing pea epicotyls. Developmental Biol 30:42.
Toppan A, Roby D, Esquerré-Tugayé MT (1982). Cell surfaces in plant-microorganism interactions. III. *In vivo* effect of ethylene on hydroxyproline-rich glycoprotein accumulation in the cell wall of diseased palnts. Plant Physiol 70:82.
Toppan A, Esquerré-Tugayé MT (1983). Cell surfaces in plant-microorganism interactions. IV. Fungal glycopeptides which elicit the synthesis of ethylene in plants. Plant Physiol (submitted).
Stuart DA, Varner JE (1980). Purification and characterization of a salt-extractable hydroxyproline-rich glycoprotein from aerated carrot discs. Plant Physiol 66:787.
Sequeira L, Graham TL (1977). Agglutination of avirulent strains of *Pseudomonas solanacearum* by potato lectin. Physiol Plant Pathol 11:43.
Van Holst GJ, Klis FM (1981). Hydroxyproline glycosides in secretory arabinogalactan protein of *Phaseolus vulgaris L.* Plant Physiol 68:979.
Varner JE (1982). Hydroxyproline-rich glycoproteins extracted from the cell walls of aerated carrot root slices. NATO-FEBS Structure and function of plant genomes (in press).

Chemical Taxonomy, Molecular Biology, and Function
of Plant Lectins, pages 187–196

FUNGAL SPORES ARE AGGLUTINATED BY PROTEINS AND ADHERE TO BEAN HYPOCOTYLS DUE TO NONSPECIFIC BINDING

Heinrich Kauss and David H. Young

Biology Department of the University
Postfach 3049
D-6750 Kaiserslautern, G.F.R.

In the last decade we have tested extracts from a variety of nonseed plant tissues and subcellular compartments for the presence of agglutinating substances. We hoped to correlate physiological parameters with lectin binding specifity and thus establish an experimental system which would enable conclusions to be drawn regarding the function of the carbohydrate binding proteins in plants. We wish to discuss here some experiments which often produced disappointing results in this respect. However, these experiments reveal a practical side of lectin research, and hopefully will help other researchers to recognize some of the pitfalls which can be encountered. We also wish to demonstrate some difficulties involved in defining precisely what is meant by a "lectin".

During our studies we often had to determine whether an agglutination was caused by classical lectins with binding sites specific for well defined small carbohydrate groups or whether the effect was nonspecific. Textbooks and reviews depict a binding site as a crypt in a protein into which a haptenic sugar fits in a complementary fashion such that all relevant moieties approach countergroups in the protein and establish a maximum degree of polar and nonpolar interactions. These pictures evolve from the convenient technique in which low molecular weight saccharides are tested for their ability to inhibit the agglutination or precipitation. This experimental approach has been of great practical value in studying many lectins such as concanvalin A (Con A), *Ricinus communis* agglutinin (RCA) or soy bean agglutinin (SBA), in which the binding site can apparently be adequately filled by monomeric sugars. However, the search for a hapten becomes more tedious when the smallest inhibitory saccharide is an oligomer built from different monomers linked in a specific manner. In such cases the appropriate saccharide is unlikely to be found by chance. If it is part of a glycoprotein and can be removed by proteolysis, it may be possible to isolate it from the assay cells. An example of a lectin which exhibits more complicated and extended binding sites is the *Phaseolus vulgaris* agglutinin (PHA) whose E-subunits preferentially bind to internal sequences of →D-Gal→D-GlcNAc→D-Man→ (Kornfeld, Kornfeld 1970) whereas the L-subunits bind to →D-Man→D-GlcNAc→D-GlcNAc→ (Toyoshima et al. 1972). Another example is the *Agaricus bisporus* lectin (Presant, Kornfeld 1972) which can be inhibited by the sequence →D-Gal→D-GalNAc→. For both lectins, however, additional sugars are also important and the respective peptide backbones of the glycopeptide significantly increase their binding affinity. These and other observations suggest that lectin binding sites can be rather large and their interactions rather complex. Even in the apparently straightforward case of Con A it has been suggested that hydrophobic interactions are involved in binding and that the binding site is not restricted to the area which interacts with α-D-glucose (Ochoa et al. 1981).

The examples above are cited to demonstrate some typical problems which can confront researchers trying to characterize interactions involving novel lectins from unusual plant extracts. Over the years we, and others, have found agglutinating activity for example in detergent extracts from various membranes (Bowles, Kauss 1976) or exposed on *Chlamydomonas* flagellar membranes (Köhle et al. 1980), but have been unsuccessful in characterizing these lectins further. There are a number of possible reasons, such as the difficulty in obtaining sufficient amounts of the source material, their instability, the simultaneous presence of several different lectins or nonspecific binding (see below). We also experienced problems in characterizing agglutinins in phloem saps, and eventually found that they contain a number of lectins, some of which could be separated and were shown to be inhibited by D-GalNAc. In addition, the phloem sap of many plants contains lectins with different molecular properties but which are all specific for 1,4-ß-GlcNAc-oligomers (for references see Kauss 1981).

Our research on an agglutinin which we isolated from

mung bean hypocotyl cell walls (Kauss, Bowles 1976), and which was found to be inhibited by D-galactose and p-nitrophenyl α-D-galactoside, was less satisfying since it was later found (Haaß et al. 1981) to be similar to a lectin, which has been purified from seeds of the same plant (Hankins, Shannon 1978). This lectin also exhibits α-galactosidase activity; it is capable of splitting p-nitrophenyl α-D-galactoside or carbohydrate groups at the surface of erythrocytes but does not attack raffinose or stachyose. The protein appears to be a member of a family of immunologically related substrate-specific α-galactosidases, found in legumes, which in only a few cases also exhibit hemagglutinating activity (Hankins et al. 1980). This raises the question of whether such agglutinins should be classified as lectins. On the other hand, it is possible that many "lectins" are actually enzymes which either have lost their activity or for which we do not know the proper substrate.

In the literature one can find many examples of agglutinating substances for which the name lectin clearly is inappropriate; some of these are listed in Table 1.

Table 1
Some examples of "non-specific" interaction which cause agglutination or precipitation. For discussion see text.

agglutinin	cells or indicator	reference
cross-linked lysozyme.	red blood cells	Hořejší, Kocourek, 1974
poly-L-lysine.	red blood cells, bacteria, virus.	Katchalski et al. 1964
chitosan	soy bean cells.	Köhle, Kauss, (unpublished)
lipopolysaccharides, lipoteichoic acid.	various animal cells.	Wicken, Knox, 1981
arabinogalactan-protein.	copolymers of 4-amino-phenyl-ß-glycosides.	Lamport, Catt, 1981
arabinoxylan.	red blood cells.	Minetti et al., 1976
various proteins and glycoproteins.	fungal spores and cell wall fragments.	Young, Kauss, 1982

Chemical cross-linking with glutaraldehyde of the univalent lysozyme results in a hemagglutinin. Its action, however, is not inhibited by D-GlcNAc-oligomers and depends on ionic strength (J. Kocourek, personal communication). It therefore appears possible that lysozyme, a very basic protein, might cause the agglutination in a nonspecific way, which does not involve its carbohydrate-specific site but depends on its ability, like other poly-cations, to bind to oppositely charged surfaces.

Clear-cut examples for such agglutination phenomena, resulting from purely ionic interactions between polycations and negatively charged surfaces, are those mediated by poly-L-lysine and chitosan (Table 1). The effect of poly-L-lysine is greatly enhanced with increasing molecular weight and results not only in agglutination but in membrane damage. At intermediate concentrations, at which cell viability is unaffected, the polycation-induced efflux of various cell constituents from suspension-cultured soy bean cells (Young et al. 1982) is followed by changes in metabolic activity. We have shown enhanced biosynthesis of cell wall polyphenols, the phytoalexin glyceollin and callose (Köhle et al. 1983). All these effects are inhibited by polyanions such as poly-L-aspartic and polygalacturonic acids, presumably, because these substances form complexes with the polycations (Young et al. 1982). Their ability to affect metabolic processes is another characteristic which these polycationic agglutinins have in common with some classical lectins, which for example can exert a mitogenic effect on lymphocytes.

This unusual analogy can be extended to include bacterial lipopolysaccharides and other bacterial surface amphiphiles which bind nonspecifically to the surfaces of a variety of eucaryotic cells (Table 1). Although the polar groups in these polymers contribute ionic binding and hydrogen bonding, there is also good evidence that their hydrophobic parts interact with hydrophobic molecules exposed at the surfaces of the target cells (Wicken, Knox 1981). The biological effects induced by the bacterial amphiphiles include those known for lectins, for example hemagglutination, mitosis or macrophage stimulation. Participation of hydrophobic interaction has also been suggested with respect to the precipitation of arabinogalactan-proteins ("ß-lectins"), which are present in many plant cell walls, by polymerized 4-aminophenyl-ß-glycosides (Table 1). This interaction occurs with the ß-anomers of various sugars, but these must be linked to the hydrophobic 4-amino-phenyl group. The hemagglutinating action of an almost protein-free arabinoxylan isolated from wheat flour (Table 1) has been suggested to result from interaction of the polysaccharide with the phospholipid bilayer of the membrane (Minetti et al. 1978). In view of the above examples it would not be surprising if such a nonspecific binding of polysaccharides also affected membrane permeability followed by alterations of the metabolism of the respective cells.

The last example of nonspecific agglutination presented in Table 1 is from own studies undertaken with the aim of finding experimental evidence for the popular hypothesis that carbohydrate-binding proteins might be responsible for the "recognition" step in plant-pathogen interactions (for references see Kauss 1981; Young, Kauss 1982). The approach we used was to test bean extracts for the presence of agglutinins specific for cell wall fragments or spores of the fungus *Colletotrichum lindemuthianum*, a pathogen of beans. We found that extracts from hypocotyls and seeds did contain agglutinins. However no correlation between agglutination titre and the resistance or susceptibility of bean varieties to different races of the fungus was evident. Furthermore, during purification of the pronase-sensitive agglutinins from seeds (Table 2) the increase in specific activity was relatively small in relation to the large losses in protein.

In addition, many fractions accumulating in the course of the purification steps contained agglutinating activity. This suggested that many different components were capable of causing the agglutination. Nevertheless, we purified two agglutinating glycoproteins to apparent purity (Table 2). During an unsuccessful search for inhibitory saccharides we also tested some commercial glycoproteins for possible inhibition and discovered that they actually caused agglutination of the spores and fungal wall fragments (Table 3). In addition some conventional proteins were effective in the same concentration range as the bean glycoproteins. These observations suggest that the agglutination does not result from group-specific binding but is probably due to a more general protein property and might involve a large surface area of the protein and multiple weak interactions. Although disappointing from the view point of lectin research, such interactions might nevertheless be of some biological significance. It has been proposed for animal cells that many recognition and adhesion phenomena might result from non-specific interactions between surface molecules (Edwards 1978). This might equally well apply to plants.

Table 2
Purification from seeds of *Phaseolus vulgaris* cv. Princor of glycoproteins which agglutinate mycelial cell wall fragments from *Colletotrichum lindemuthianum*.

Purification	Total Protein [mg]	Total Carboh. [mg]	Agglutination [au/μg protein][a]	Agglutination [au/μg carboh.][a]
Crude extract	3 206	3 801	17	14
Heat treatment	881	1 728	35	18
DEAE-Sephadex	107	30	20	72
Cibacron GA 3	41	4	22	233
1.urea gel (10 %)				
fraction 4	2	0.4	113	562
fraction 5	1	0.2	76	480
2.urea gel (5 %)[b]				
fraction 4	0.2	0.2	174	114
fraction 5	0.3	0.2	36	60

[a] au, agglutination unit, defined as the amount of agglutinin in a 50 μl aliquot which gives a titre of 1.

[b] Fraction 4 and 5 from the first gel electrophoresis were rerun separately on 5 % gels. Each sample gave a single but different glycoprotein band which corresponded with agglutinating activity.
Data taken from Young and Kauss (1982) with permission of Academic Press Inc. (London) Ltd.

We have found evidence for such a phenomenon in a study of the adhesion of *Colletotrichum* spores to bean plants (Table 4). Using an assay in which hypocotyl segment were rolled in a spore suspension, it was found that a significant proportion of the spores adhered, and that this could be prevented by various proteins and glycoproteins, which were shown to bind to the spores by means of nonspecific interactions (Table 3). The ability of the detergent Tween 80 to inhibit spore adhesion to the plant surface suggested that hydrophobic interactions might be involved. Hypocotyls which were dipped 3 times for 1 sec in chloroform to remove surface wax also showed greatly reduced adhesion (data not shown), consistent with the possible involvement of hydrophobic groups on the plant surface in adhesion.

Table 3

Agglutination of mycelial cell wall fragments and spores from *Colletotrichum lindemuthianum* by various proteins and glycoproteins.

Protein	Minimum concentration required [μg/ml] Cell wall	Spores
Glycoprotein (fraction 4)[a]	0.5	–[b]
Glycoprotein (fraction 5)[a]	0.1	–[b]
Bovine serum albumin	0.5	0.5
Ovalbumin	0.5	1.0
Fetuin	2	0.25
Cytochrome c	2	4
Haemoglobin	4	4
Thyroglobulin	31	31
γ-Globulin	31	62
Protamine sulphate	250	500

[a] The two glycoproteins were isolated from *Phaseolus vulgaris* seeds by the procedure shown in Table 2.
[b] Agglutination of spores occured at concentrations similar to that necessary for cell walls, but was not tested for this spore preparation.
Data taken from Young and Kauss (1982) with permission of Academic Press Inc. (London) Ltd.

In addition, adhesion appears to be partly inhibited by respiratory inhibitors like antimycin A and sodium azide. This might either indicate turnover of relevant molecules on the spore surface or the requirement of an undisturbed membrane potential.

Evidence that spore adhesion to the plant surface was due to nonspecific binding was also obtained in further experiments in which we examined adhesion to the surface of polystyrene microtitre plate wells (Table 4). The ability to adhere to polystyrene has been used previously to screen bacteria for surface hydrophobicity (Rosenberg 1981). We found that spore adhesion to polystyrene could also be inhibited by proteins, glycoproteins and Tween 80 (Table 4), as well as by the aforementioned metabolic inhibitors.

Table 4
Inhibition by various substances of the adhesion of *Colletotrichum lindemuthianum* spores.

Substance added	% spores adhering to bean hypocotyls[a]	polystyrene surface[b]
None (control)	100	100
Bovine serum albumin	0	2
γ-Globulin	0	3
Fetuin	3	3
Haemoglobin	12	3
Cytochrom c	26	0
Ovalbumin	28	11
Tween 80	14	8

[a] Assay: Two bean (cv. Simplobel) hypocotyl segments (2 cm) were rolled in 9 x 10^5 spores (race kappa) in 3 ml 10 mM citrate-phosphate, pH 5.0, for 1 h and the unbound spores counted with a coulter counter. Substances were added at 100 µg/ml. In the control about one third of the added spores adhered to the hypocotyl.

[b] Assay: 4 x 10^5 spores in 0.4 ml of the same buffer were allowed to settle for 1 h in flat-bottomed polystyrene microtitre wells. Before and after removing the supernatant and 2 washings with the buffer, the spores per 0.8 mm^2 were counted using a Zeiss inverted microscope. In the control 2300 out of 6700 spores per mm^2 adhered to the polystyrene. Substances were added at 10 µg/ml. Different batches of spores were used in the two assays.

Methods and more results will be published elsewhere.

In both assays more spores adhered in the presence of dilute buffer (10 mM citrate-phosphate, pH 5) or added salts (10^{-4}-10^{-2} M NaCl) than in deionized water alone. This may reflect the known ability of added solutes to enhance hydrophobic interactions (Tanford 1980). Proteins had to be added at a higher concentration in the hypocotyl assay than in the microtitre plate assay to obtain a similar effect (Table 4). A possible explanation is that in the hypocotyl assay the suspension was constantly rolled, whereas in the microtitre assay the spores were allowed to settle without disturbance; thus the mechanical forces applied in the two assays were quite different. Although it is difficult to compare adhesion quantitatively in two such different assays, it would appear that the same nonspecific mechanism operates in both cases.

References:

Bowles DJ, Kauss H (1976) Characterization, enzymatic and lectin properties of isolated membranes from *Phaseolus aureus*. Biochem. Biophys. Acta 443:360-374.

Edwards PAW (1978) Differential cell adhesion may result from nonspecific interactions between cell surface glycoproteins. Nature 271:248.

Haaß D, Frey R, Thiesen M, Kauss H (1981) Partial purification of a hemagglutinin associated with the cell walls from hypocotyls of *Vigna radiata*. Planta 151:490.

Hankins CN, Shannon LM (1978) The physical and enzymatic properties of a phytohemagglutinin from mung beans. J Biol Chem 253:7791-7797.

Hankins CN, Kindinger JI, Shannon LM (1980) Legume α-galactosidases which have hemagglutinin properties. Plant Physiol 65:618-622.

Horejsi V, Kocourek J (1974) The covalent oligomers of lysozyme:first case of semisynthetic hemagglutinins. Experientia 30:1348.

Katschalski E, Sela M, Silman HI, Berger A (1964) Polyamino acids as protein models. In Neurath H (ed): "The proteins." Sec. ed. Vol II. Academic Press New York London, p 405.

Kauss H (1981) Lectins and their physiological role in slime molds and in higher plants. In Tanner W, Loewus FA (eds): "Plant carbohydrates II." Springer-Verlag Berlin Heidelberg New York, p 627.

Kauss H, Bowles D (1976) Some properties of carbohydrate-binding proteins (lectins) solubilized from cell walls of *Phaseolus aureus*. Planta 130:169-174.

Köhle D, Lang W, Kauss H (1980) Agglutination and glycosyltransferase activity of isolated gametic flagella from *Chlamydomonas reinhardii*. Arch Microbiol 127:239.

Köhle H, Young DH, Kauss H (1983) Biochemical and physiological changes in suspension-cultured soybean cells elicited by treatment with chitosan. Plant Science Letters, in press.

Kornfeld R, Kornfeld S (1970) The structure of a phytohemagglutination receptor site from human erythrocytes. J Biol Chem 245:2536.

Lamport DTA, Catt JW (1981) Glycoproteins and enzymes of the cell wall. In Tanner W, Loewus FA (eds): "Plant carbohydrates II." Springer-Verlag Berlin Heidelberg New York, p 133.

Minetti M, Aducci P, Teichner A (1976) A new agglutinating activity from wheat flour inhibited by tryptophan. Biochim Biophys Acta 437:505.

Minetti M, Teichner A, Aducci P (1978) Interaction of neutral polysaccharides with human erythrocyte membrane: involvement of phospholipid bilayer. Biochem Biophys Res Commun 80:46.

O'Brian RL, Parker JW, Dixon JFP (1978) Mechanisms of lymphocyte transformation. Progr Mol Subcell Biol 6:201.

Ochoa JL, Sierra A, Córdoba F (1981) On the specifity and hydrophobicity of lectins. In Bøg-Hansen TC (ed): "Lectins - biology, biochemistry, clinical biochemistry." Vol 1, Walter de Gruyter Berlin New York, p 73.

Presant CA, Kornfeld S (1972) Characterization of the cell surface receptor for the *Agaricus bisporus* hemagglutinin. J Biol Chem 247:6937.

Rosenberg M (1981) Bacterial adherence to polystyrene: A replica method of screening for bacterial hydrophobicity. Applied and Environmental Microbiology 42:375

Tanford C (1980) The hydrophobic effect:"Formation of micelles and biological membranes."John Wiley New York.

Toyoshima S, Funkuda M, Osawa T (1972) Chemical nature of the receptor site for various phytomitogens. Biochemistry 11:4000.

Wicken AJ, Knox KW (1981) Chemical composition and properties of amphiphiles. In Shockman GD, Wicken AJ (eds): "Chemistry and biological activities of bacterial surface amphiphiles" Academic Press, New York London, p 1.

Young DH, Kauss H (1982) Agglutination of mycelial cell wall fragments and spores of *Colletotrichum lindemuthianum* by various proteins. Physiol Plant Pathol 20:285.

Young DH, Köhle H, Kauss H (1982) Effect of chitosan on membrane permeability of suspension-cultured *Glycine max* and *Phaseolus vulgaris* cells. Plant Physiol. 70:1449.

Chemical Taxonomy, Molecular Biology, and Function of Plant Lectins, pages 197–207

DISTRIBUTION AND PROPERTIES OF THE *DOLICHOS BIFLORUS* LECTINS: A MODEL SYSTEM FOR EXPLORING THE ROLE OF LECTINS IN PLANTS

Marilynn E. Etzler

Department of Biochemistry & Biophysics
University of California
Davis, CA 95616

INTRODUCTION

Lectins are particularly abundant in legumes where they constitute up to 10% of the soluble protein of the seed (Liener 1976). In addition to their presence in the seeds, lectins have been found in other tissues of this family of plants, although the relationship of these lectins to the seed lectins in many cases is not yet clear (for review see Etzler, in press). This abundance and distribution of lectins in the tissues of legumes suggests that these molecules may serve one or more important functions during the life cycle of the plant. Yet, although many such roles have been proposed, at present we have little evidence to confirm any of these roles as a biological function of lectins in these plants (for review see Etzler, in press).

A variety of approaches have been utilized in an effort to obtain information on the function of lectins. In our laboratory we have chosen the *Dolichos biflorus* plant as an experimental system and are combining studies on the molecular properties of the lectins from this plant with studies on the distribution of these lectins during the life cycle of the plant and the regulation of their production. It is hoped that such a concerted study at the molecular, subcellular and systemic levels may provide basic information that will eventually yield clues as to the function(s) of these molecules in the plant.

MOLECULAR PROPERTIES OF DOLICHOS BIFLORUS LECTINS

The seeds of the Dolichos biflorus plant contain a lectin that agglutinates type A erythrocytes and precipitates blood group A substance (Bird 1952; Boyd, Shapleigh 1954) due to its specificity for terminal nonreducing α-linked N-acetylgalactosamine residues (Etzler, Kabat 1970). This lectin is a glycoprotein that has a molecular weight of about 110,000 and is a tetramer consisting of apparently equal amounts of two types of subunits, I and II (Carter, Etzler 1975a). Equilibrium dialysis studies have shown that it has two carbohydrate combining sites per molecule and has an association constant of 3.2×10^3 liters/mole for methyl α-N-acetylgalactosamine (Etzler, et al. 1981). The lectin is heterogeneous with respect to its carbohydrate content and can be subdivided into several fractions by chromatography on concanavalin A-Sepharose (Carter, Etzler 1975a). These fractions are almost identical with respect to their carbohydrate binding activities and specificities (Carter, Etzler 1975a; Borrebaeck, Etzler 1980; Etzler, et al. 1981).

Subunits I and II of the seed lectin have similar amino acid and carbohydrate compositions and give reactions of identity when tested in immunodiffusion against antisera prepared against either the seed lectin or its subunits (Carter, Etzler 1975b). Both subunits have identical amino terminal amino acid sequences (Etzler, et al. 1977), and studies of peptides obtained after cleavage of the subunits with cyanogen bromide or the V-8 protease from Staphylococcus aureus show that all peptides except the COOH-terminal peptides are common to both subunits (Carter, Etzler 1975c; Roberts, et al. 1982). Subunit I has a slightly larger molecular weight than Subunit II, and studies with carboxypeptidases have shown that the subunits differ in their COOH-terminal amino acids (Carter, Etzler 1975b; Roberts, et al. 1982). These studies suggest that both subunits of the lectin may represent different degrees of completion or posttranslational modification of a common polypeptide chain and thus be products of the same gene.

Despite the structural similarities between the seed lectin subunits, only Subunit I has been found to have carbohydrate binding activity. This subunit has an association constant for N-acetylgalactosamine similar to that of the intact lectin and may be primarily responsible for the carbohydrate binding properties of the lectin (Etzler, et al. 1981). A monoclonal antibody specific for Subunit I has been produced; this antibody inhibits the activity of the lectin, and the ability of this antibody to combine with the lectin is specifically inhibited by the presence of N-acetylgalactosamine (Borrebaeck, Etzler 1981). These results indicate that the monoclonal antibody may be directed against a determinant at the active site of the lectin or at a site that is conformationally interdependent with the active site. Further comparative studies of these two lectin subunits should be useful in determining the relationship of structure to function of this molecule.

The stems and leaves of this plant contain a glycoprotein that cross reacts with antibodies to the seed lectin (Talbot, Etzler 1978a). This cross reactive material (CRM) is similar to the seed lectin in its amino acid and carbohydrate composition and is a dimer consisting of two types of subunits. One of these subunits has an electrophoretic mobility in SDS-urea polyacrylamide gels identical to Subunit I of the seed lectin; the other subunit has a higher molecular weight than either of the seed lectin subunits (Talbot, Etzler 1978b). An electrophoretic comparison of these subunits is shown in Figure 1. Both subunits have identical amino terminal amino acid sequences and these sequences are identical to the sequences of the seed lectin subunits with the exception of an aspartate in place of an asparagine at residue 2 (Talbot, Etzler 1978b). A comparison of peptide maps generated by chymotrypsin cleavage of the CRM and the seed lectin indicates that most peptides are common to both molecules (Jin, et al. 1983). These findings indicate that both the CRM and the seed lectin subunits may represent modifications of a common polypeptide chain.

When tested in the presence of physiological saline, the CRM has not been found to have any hemagglutinating or precipitating activity nor does it inhibit the agglutination of erythrocytes by the seed lectin (Talbot, Etzler 1978b). At low ionic strength, however, the CRM binds to hog blood group A+H substance and this binding is inhibited by both N-acetylglucosamine and N-acetylgalactosamine (Etzler, Borrebaeck 1980). The CRM has also become activated under conditions that have not

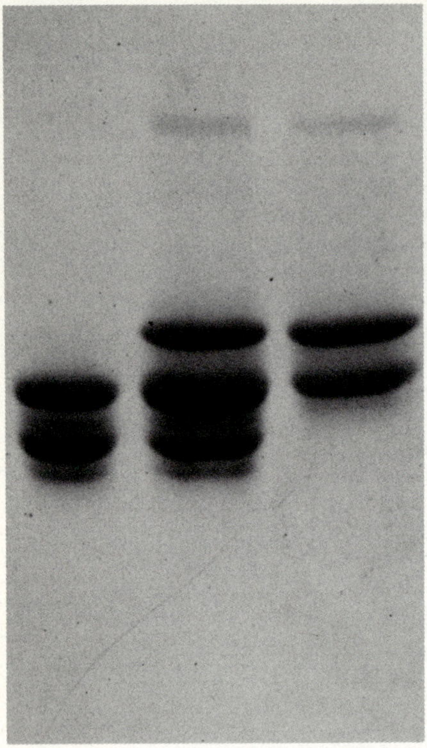

Fig. 1. Discontinuous polyacrylamide electrophoresis of isolated seed lectin (left), CRM (right) and a mixture of these two proteins (center) on pH 9.7 glycine gels in the presence of 0.1% sodium dodecyl sulfate and 8.0 M urea.

yet been defined to form an active species with hemagglutinating activity much greater than that of the seed lectin; this activation causes no apparent change in its subunit content (unpublished results). The CRM itself may therefore be considered as a lectin; its ability to exist in active and inactive states indicates that its activity might be under some type of regulatory control in the plant. A summary of the properties of the seed lectin and CRM is given in Table 1.

In addition to the seed lectin and the CRM, another carbohydrate binding protein has also been found in the Dolichos biflorus plant. This glycoprotein, which is present in the roots as well as the stems and leaves, was isolated by affinity chromatography on blood group A+H substance (Quinn, Etzler 1983). It has a different composition from the seed lectin and CRM and does not agglutinate erythrocytes. Whether this glycoprotein is

TABLE 1. MOLECULAR PROPERTIES OF DOLICHOS BIFLORUS LECTINS

	Seed Lectin	Stem and Leaf Lectin (CRM)
Composition	Rich in aspartate and serine; no cysteine About 5% carbohydrate	Rich in aspartate and serine; no cysteine About 5% carbohydrate
Molecular weight	110,000	70,000
Subunit Structure	Tetramer (I_2II_2)	Dimer (I)U
Specificity	α-GalNAc	(GlcNAc, GalNAc)

a lectin has yet to be established; studies on its molecular structure are in progress in our laboratory.

DISTRIBUTION OF LECTINS IN DOLICHOS BIFLORUS PLANT

The distributions of both the seed lectin and the CRM in the Dolichos biflorus plant have been studied using radioimmunoassays (Talbot, Etzler 1978a; Roberts, Etzler 1980). The seed lectin is localized in the cotyledons where it constitutes up to 10% of the soluble protein of this tissue. It appears at about the 27th day after flowering and within two days reaches the levels found in the mature seeds. As the seeds germinate, the level of the lectin decreases at about the same rate as the level level of other proteins in the cotyledons (Talbot, Etzler 1978a). Immunofluorescence, immunocytochemical and cell fractionation studies have shown that a major portion of the seed lectin is associated with the protein bodies where it is present at the periphery of these organelles. Some lectin has also been found in the cytoplasm and at the periphery of the starch granules (unpublished results). The seed lectin has not been found in other vegetative parts of the plant.

In contrast to the seed lectin, no significant amounts of CRM were detected in dormant seeds nor in the first day after imbibition of the seeds with water. CRM appears at the second day of germination and reaches a maximum level by the fourth day; measurements of the amount of CRM/mg protein extracted from the plants slightly decrease after this point as shown in Figure 2.

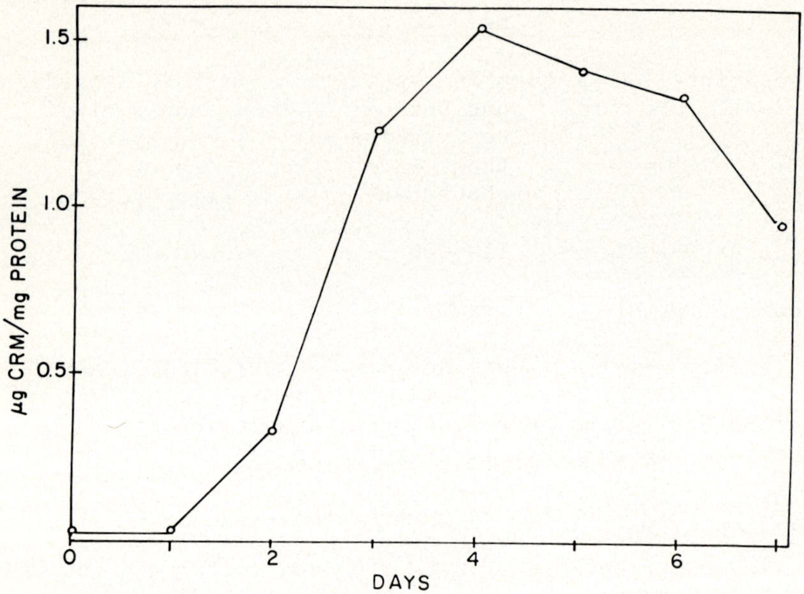

Fig. 2. CRM content of Dolichos biflorus plants during germination and early seedling growth.

Most of the CRM was found to be present in the epicotyl and leaves of the germinating plants; low levels were found in the hypocotyls and cotyledons. The distribution of the CRM was followed in the developing plant and are found to be present in the leaves and in the stems where it was most abundant in those internodes undergoing elongation. No CRM was detected in the roots of the plant (Roberts, Etzler 1980).

The subcellular distribution of the CRM in the plant was studied by immunofluorescence, immunocytochemistry and cell fractionation studies. These studies showed that a significant portion of the CRM is associated with the cell walls of the various tissues of the stems and leaves. The CRM appeared to be noncovalently associated with the walls since it could be released from cell wall enriched fractions by treatment with high salt, detergents, and EDTA as well as by treatment with cellulase and pectinase (unpublished results).

PRODUCTION OF LECTIN IN CELL CULTURES OF DOLICHOS BIFLORUS

Studies now in progress in our laboratory have defined conditions under which CRM can be produced in callus cultures of the *Dolichos biflorus* plant. The relative amounts of auxin and cytokinin are important in allowing the expression of the CRM. Our present results indicate a direct correlation of CRM production with the greening of the tissue but further work must still be done in this area. CRM produced in callus cultures has been extracted and isolated and found to be identical to the CRM isolated from the intact plant.

CONCLUSION

The presence of at least two lectins in the *Dolichos biflorus* plant and their different distributions and subcellular localizations suggest that lectins may be utilized by legumes for more than one function. The seed lectin from this plant resembles a number of other legume seed lectins in its molecular properties (for review see Goldstein, Hayes 1978) and has been found to have homologies in primary structure with lectins from closely related legumes (Etzler, *et al*. 1977). The localization of this lectin in the protein bodies of the cotyledons correlates with work by other investigators on the subcellular localization of the seed lectins from *Phaseolus vulgaris*, *Vicia faba*, soybeans, and peas (Pusztai, *et al*. 1978; Bollini, Chrispeels 1978; Begbie 1979; Manen, Pusztai 1982; Weber, *et al*. 1978; Horisberger, Vonlanthen, 1980; Van Driessche, *et al*. 1981). Such a localization is compatible with proposed roles for lectins as storage proteins (Boyd 1963) or as aids in the packaging or mobilization of storage materials (Weber, Neuman 1980). The CRM, which is structurally related to the seed lectin, is present in the stems and leaves of the *Dolichos biflorus* plant where a significant portion of it is associated with the cell wall. A cell wall associated lectin has also been reported in mung bean hypocotyls where it was suggested that it might be involved in cell wall extension (Kauss, Glaser 1974; Haasz, *et al*. 1981). The relationship of the mung bean cell wall lectin to a seed lectin from this plant has not been established.

The similarities in molecular properties of the seed lectin and CRM suggest that these molecules may either be products of genes that arose by duplication or may represent different posttranslational modifications of a common gene product. Studies now in progress in our laboratory at the translational and transcriptional levels may help to distinguish between these possibilities. The finding that the CRM can exist in both inactive as well as active states raises the intriguing possibility that the activity of this lectin may be under some type of regulatory control. Such a possibility can only be determined by further structural work on this molecule.

Our recent establishment of conditions that support the appearance of the CRM in cultures of *Dolichos biflorus* enables us to proceed with studies aimed at determining the nature of regulation of the production of this lectin. Meimeth *et al*. (1982) have also recently reported the production of a lectin in callus cultures derived from *Psophocarpus tetragonolobus*.

By focusing on the *Dolichos biflorus* plant and integrating studies at the molecular, cellular and systemic levels we have developed a model system for comparing lectins produced by different tissues and the regulation of their production. It is hoped that basic information gained in this study may eventually be useful in elucidating the function of lectins in plants.

REFERENCES

Begbie R (1979). A non-aqueous method for the subcellular fractionation of cotyledons from dormant seeds of *Phaseolus vulgaris* L. Planta 147:103.

Bird GWG (1952). Relationship of the blood sub-groups A_1, A_2, and A_1B, A_BB to haemagglutinins present in the seeds of *Dolichos biflorus*. Nature 170:674.

Bollini R, Chrispeels MJ (1978). Characterization and subcellular localization of vicilin and phytohemagglutinin, the two major reserve proteins of *Phaseolus vulgaris* L. Planta 142:291.

Borrebaeck C, Etzler ME (1980). A study of the binding properties of the isolectins from the *Dolichos biflorus* plant using affinity electrophoresis. FEBS Lett 117:237.

Borrebaeck CAK, Etzler ME (1981). Production and characterization of a monoclonal antibody against the seed lectin of the Dolichos biflorus plant. J Biol Chem 256:4723.

Boyd WC (1963). The lectins: their present status. Vox Sang 8:1.

Boyd WC, Shapleigh E (1954). Specific precipitating activity of plant agglutinins (lectins). Science 119:419.

Carter WG, Etzler ME (1975a). Isolation, characterization and subunit structures of multiple forms of Dolichos biflorus lectin. J Biol Chem 250:2756.

Carter WG, Etzler ME (1975b). Isolation and characterization of subunits from the predominant form of Dolichos biflorus lectin. Biochemistry 14:2685.

Carter WG, Etzler ME (1975c). Isolation and characterization of cyanogen bromide fragments and a glycopeptide from the Dolichos lectin. Biochemistry 14:5118.

Etzler ME. Function of plant lectins. In Liener IE, Sharon N, Goldstein IJ (eds): "The Chemistry and Biology of Lectins," New York: Academic Press Inc, in press.

Etzler ME, Borrebaeck C (1980). Carbohydrate binding activity of a lectin-like glycoprotein from stems and leaves of Dolichos biflorus. Biochem Biophys Res Comm 96:92.

Etzler ME, Kabat EA (1970). Purification and characterization of a lectin (plant hemagglutinin) with blood group A specificity from Dolichos biflorus. Biochemistry 9:869.

Etzler ME, Gupta S, Borrebaeck C (1981). Carbohydrate binding properties of the Dolichos biflorus lectin and its subunits. J Biol Chem 256:2367.

Etzler ME, Talbot CF, Ziaya P (1977). NH_2-terminal sequences of the subunits of Dolichos biflorus lectin. FEBS Lett 82:39.

Goldstein IJ, Hayes CE (1978). The lectins: carbohydrate-binding proteins of plants and animals. Adv Carbohydr Chem Biochem 35:127.

Haasz D, Frey R, Thiesen M, Kauss H (1981). Partial purification of hemagglutinin associated with cell walls from hypocotyls of Vigna radiata. Planta 151:490.

Horisberger M, Vonlanthen M (1980). Ultrastructural localization of soybean agglutinin on thin sections of Glycine max (Soybean) var. Altona by the gold method. Histochem 65:181.

Jin Y-J, Roberts DM, Etzler ME (1983). Comparison of the peptide maps of two lectins from the Dolichos biflorus plant. Fed Proc 42:2015.

Kauss H, Glaser C (1974). Carbohydrate-binding proteins from plant cell walls and their possible involvement in extension growth. FEBS Lett 45:304.

Liener IE (1976). Phytohemagglutinins (phytolectins). Ann Rev Plant Physiol 27:291.

Manen JF, Pusztai A (1982). Immunocytochemical localisation of lectins in cells of Phaseolus vulgaris L. seeds. Planta 155:328.

Meimeth T, Thanh Van KT, Marcotte J-L, Trinn TH, Clarke AE (1982). Distribution of lectins in tissues, derived callus and roots of Psophocarpus tetragonolobus (Winged bean). Plant Physiol 70:579.

Pusztai A, Stewart JC, Watt WB (1978). A novel method for the preparation of protein bodies by filtration in high (over 70% w/v) sucrose-containing media. Plant Sci Lett 12:9.

Quinn JM, Etzler ME (1983). A carbohydrate binding protein from the roots of Dolichos biflorus that differs from the seed lectin. Fed Proc 42:2015.

Roberts DM, Etzler ME (1980). Developmental study of a protein that cross reacts with antibodies to the Dolichos biflorus seed lectin. Plant Physiol 66 (Suppl):51.

Roberts DM, Walker J, Etzler ME (1982). A structural comparison of the subunits of the Dolichos biflorus seed lectin by peptide mapping and carboxyl terminal amino acid analysis. Arch Biochem Biophys 218:213.

Talbot CF, Etzler ME (1978a). Development and distribution of Dolichos biflorus lectin as measured by radioimmunoassay. Plant Physiol 61:847.

Talbot CF, Etzler ME (1978b). Isolation and characterization of a protein from leaves and stems of Dolichos biflorus that cross reacts with antibodies to the seed lectin. Biochemistry 17:1474.

Van Driessche E, Smets G, Dejaegere R, Kanarek L (1981). The immuno-histochemical localization of lectin in pea seeds (Pisum sativum L.). Planta 153:287.

Weber E, Neumann D (1980). Protein bodies, storage organelles in plant seeds. Biochem Physiol Pflanzen 175:279.

Weber E, Manteuffel R, Neumann D (1978). Isolation and characterization of protein bodies of Vicia faba seeds. Biochem Physiol Pflanzen 172:597.

Chemical Taxonomy, Molecular Biology, and Function
of Plant Lectins, pages 209–223

Trifoliin A: a Rhizobium-Recognition Lectin in White Clover Roots

Frank B. Dazzo and John E. Sherwood

Department of Microbiology & Public Health
Michigan State University
East Lansing, MI 48824

Rhizobium is a genus of gram-negative soil bacteria which can infect and nodulate legume roots, forming a nitrogen fixing symbiosis. There is a high degree of host-specificity in this symbiosis, e.g., R. meliloti nodulates alfalfa but not clover, and R. trifolii nodulates clover but not alfalfa. Therefore, this symbiosis can be used to study plant-microbe interactions as models of cellular recognition.

Over the last decade, much work has focused on the basic mechanisms which lead to the successful infection of legume roots by the nitrogen-fixing symbiont. A hallmark of this effort has been the discovery that certain legume lectins fulfill many criteria of a cell recognition molecule in this plant-bacterial symbiosis. Since this subject has been exhaustively reviewed elsewhere (Schmidt, 1979; Bauer 1980; Dazzo 1981; Dazzo, Truchet 1983), we will only highlight some important developments which illustrate the complexity of this cellular recognition phenomenon and point to future research priorities.

The focus of lectin involvement in the Rhizobium-legume symbiosis began with the demonstration that bean lectin interacts with bean rhizobia (Hamblin, Kent 1973). A year later, Bohlool & Schmidt (1974) reported that the soybean-nodulating rhizobia interact specifically with soybean lectin from seeds. This specific correlation was subsequently confirmed (Bhuvaneswari et al. 1976) and extended to include several other legume-Rhizobium symbioses (Dazzo, Hubbell 1975b; Wolpert, Albersheim 1976;

Kamberger 1979a; Paau et al. 1980; Kato et al. 1980). This specificity became the basis for formulating the "lectin recognition hypothesis": that recognition between rhizobia and the legume host root involves a binding of the plant lectins to unique carbohydrates found exclusively on the bacterial symbiont as a prelude to infection. However, as early as 1975, it became clear that not all legume seed lectins specifically recognize the corresponding rhizobial symbiont. For example, concanavalin A from jackbean seeds bound in a hapten-reversible manner to rhizobia which do not nodulate jackbean roots (Dazzo, Hubbell 1975a), and soybean lectin from soybean seeds bound to some peanut rhizobia which do not nodulate soybean roots (Pueppke et al, 1980). These results emphasize that tests of the lectin recognition hypothesis should not be prejudiced to only consider the classical hemagglutinating lectins from legume seeds, but rather should focus on whether root lectins bind specifically to the rhizobial symbiont before infection.

We directed our attention to the *Rhizobium*-clover symbiosis to pursue the lectin recognition hypothesis. Clover is admirably suited to studies of the infection process due to its small size and blue autofluorescence of its roots which allow detailed quantitative light and fluorescence microscopic studies. In addition, its bacterial symbiont, *R. trifolii*, is easily cultivated in simple, defined culture media.

Transmission electron microscopy of *R. trifolii* shows a dense, fibrillar capsule completely surrounding the cell (Dazzo, Brill 1979). Recent evidence indicates that the capsule develops first at one pole, then extends around the sides of the cell length, and finally covers the opposite pole (Truchet, Sherwood, Dazzo, and Vasse, manuscript in preparation). The capsule consists of several polysaccharides, including both neutral glucans and acidic heteropolysaccharides. When fully encapsulated cells are inoculated on clover seedling roots at high inoculum density, they rapidly clump at the tips of the root hair cells. Root hairs are the target cells through which infection begins. During the next several hours, other bacteria attach polarly along the sides of the root hair. A very marked curling of the root hair tip is induced 1-2 days later. Occasionally, bacteria entrapped within the curl penetrate the root hair cell wall and form a tubular structure called the "infection thread", which is an unambiguous sign of successful infection. Host specificity is expressed at some event preceeding the formation of these infection threads within the root hair (Li, Hubbell 1969). Within a few days, the invading bacteria induce proliferation of cortical cells in the root which eventually emerge as a nodule. These nodules contain the microbial symbiont and reduce atmospheric nitrogen into ammonia, making it readily available to the plant.

By examining the lectin recognition hypothesis in clover, we discovered a new lectin which does not normally agglutinate erythrocytes (Dazzo et al. 1978). This lectin is a carbohydrate-binding glycoprotein with approximately 53,000 dalton subunit molecular weight (corrected from earlier estimates) and aggregates at neutral pH, near its isoelectric point, into particles which are approximately 10 nm in diameter. We have named this clover lectin "trifoliin A" (from *Trifolium*, the genus of clovers).

Trifoliin A has several unique properties which qualify it as a cell-recognition molecule in the *Rhizobium*-clover symbiosis (Dazzo et al. 1978). Trifoliin A specifically binds to and agglutinates the clover symbiont, *Rhizobium trifolii*, at microgram per ml concentrations. Its interaction with cells and their surface polysaccharides is specifically inhibited by the hapten 2-deoxyglucose (Dazzo, Brill 1977; Dazzo et al. 1978; Hrabak et al. 1981). Based on Ouchterlony immunodiffusion, the lectin from the seed is antigenically related to a lectin from the root. Immunofluorescence microscopy reveals this lectin on differentiated clover root hair cells and not on undifferentiated epidermal cells. The distribution of trifoliin A immunofluorescence on root hairs is a gradient which is greatest at the growing hair tip and least at the base of the root hair.

We examined the role of trifoliin A in the attachment of *R. trifolii* to clover root hairs. Quantitative microscopic techniques showed that rhizobia which could infect clover attached in the highest numbers to these root hair cells (Dazzo et al. 1976). At 30 mM, the hapten 2-deoxyglucose (but not the analogue 2-deoxygalactose) specifically inhibited attachment of clover rhizobia to clover root hairs (Dazzo et al. 1976; Zurkowski 1980), but did not inhibit attachment of alfalfa rhizobia to alfalfa root hairs. It is now known that an agglutinin on alfalfa root hairs specifically binds the alfalfa symbiont, R. meliloti, and 2-deoxyglucose is not an effective hapten for this alfalfa agglutinin (Paau et al. 1981). Only about 90% of the R. trifolii attachment to clover root hairs was inhibited by 2-deoxyglucose, providing evidence for multiple mechanisms of attachment -- some which are host-symbiont specific and 2-deoxyglucose inhibitable, and others which are non-specific at background levels. (The latter mechanism is 2-deoxyglucose resistant, mediated perhaps by plant polysaccharide-bacterial polysaccharide or bacterial pili-plant polysaccharide interactions). Similarly, attachment of R. japonicum and R. leguminosarum to root hairs of their respective hosts is Rhizobium-specific and largely inhibited by corresponding haptens of the host root lectins (Stacey et al. 1980; Kato et al. 1980, 1981). These studies strongly suggest that a major mechanism of rhizobial attachment to host root hairs involves the symbiont-specific interaction of host lectins and the surface polymers of the bacterial symbiont. However, specificity was not found in quantitative root attachment employing very large inocula (10^9 - 10^{10} cells per seedling) of radiolabelled rhizobia (Chen, Phillips 1976), but many unattached bacteria could have contributed to this result. Recently, van Rensberg & Strijdom (1982) demonstrated non-specific loose attachment and host-specific firm attachment of rhizobia to their respective clover and alfalfa hosts which correlated directly with success in interstrain competition for nodulation sites on the root.

We next studied the interactions between trifoliin A-binding polysaccharides of R. trifolii and the clover root surface (Dazzo, Brill 1977). The bacterial capsular polysaccharide labelled with fluorescein isothiocyanate (which fluoresces green) specifically bound to the root hair cells, matching the distribution of trifoliin A. This host-specific detection of receptor sites has been extended to include rhizobial symbioses with alfalfa, peas, and wild soybean (Hughes & Elkan 1981; Kato et al. 1980). In every case where the hapten is known, binding of the bacterial polysaccharide to the root hairs was inhibited specifically by the addition of hapten sugars of the host lectin. This result strongly suggests that host lectin on the root hairs provides recognition sites for the bacterial symbiont. Hapten sugars selectively elute the clover, pea, and soybean lectin from off the corresponding root (Dazzo, Brill 1977; Dazzo et al. 1978; Kijne et al. 1980; Gade et al. 1981), suggesting that some of the lectin is reversibly bound to receptors on the root surface. The hapten could also block exposed sites of lectin irreversibly bound to the root hair surface.

Other studies support the hypothesis that the host lectin and its saccharide receptors are involved in this initial attachment of the bacterium to the root hairs. First, certain non-infective mutants of R. trifolii and heterologous rhizobia do not bind trifoliin A and only bind to clover root hairs at background levels (Dazzo et al. 1976). Second, growth of the root in nitrate-supplemented media results in parallel reductions in levels of surface-associated trifoliin A and root hair attachment by clover rhizobia (Dazzo, Brill 1978). Third, the transient level of these bacterial polysaccharide receptors directly correlates with the quantity and orientation of bacterial attachment to the clover root surface (Dazzo et al. 1979; Truchet, Sherwood, Dazzo, and Vasse, manuscript in preparation). Fourth, intergeneric transformations of Azotobacter vinelandii with R. trifolii DNA led to the formation of hybrid cells which make clover lectin receptors (Bishop et al. 1977) and which specifically acquire the ability to attach to the clover root hairs in a hapten-inhibitable fashion (Dazzo, Brill 1979). Fifth, Fab monovalent fragments against a cross-reactive clover root antigen (Dazzo, Hubbell 1975b) compete specifically with trifoliin A for saccharide determinants on the bacterial surface and block the bacterial attachment to the clover root hairs (Dazzo, Brill 1979). This key observation and the hapten-elution data described above suggest that the bacteria and clover root have similar receptors for the multivalent lectin. Sixth, the large indigenous plasmids of clover rhizobia which confer host specificity and nodulation characteristics (Johnston et al. 1978; Hooykaas et al. 1980), also carry genetic information required for synthesis of the bacterial lectin receptor and for the ability of the bacterial cell to attach specifically to the clover root hair (Dazzo, Truchet, and Hooykaas, manuscript in preparation). Zurkowski (1980) showed that the 2-deoxyglucose sensitive step of clover root hair attachment is a phenomenon which is encoded on this large, nodulation plasmid of R.

trifolii. This opens a new horizon to examine events involved in reçognition at the root hair level which are encoded on the (Sym)biotic nodulation plasmid.

We are presently investigating the interactions between the lectin and its receptors in more detail. Transmission electron microscopy of Rhizobium trifolii in the process of attaching to the root hair shows the bacterial capsule in contact with electron-dense aggregates which lie on the outer periphery of the clover root hair cell wall (Dazzo, Hubbell 1975b). We have recently isolated these electron-dense particles which stain positively with uranyl acetate and accumulate on the root hair surface (Truchet, Sherwood, and Dazzo, manuscript in preparation). Fluorescence and electron microscopic studies indicate that these "Truchet particles" contain both trifoliin A and another component which binds this lectin. They can be dislodged from intact roots by high speed shaking or by gentler elution with the effective hapten, 2-deoxyglucose, consistent with previous hapten-elution data (Dazzo, Brill 1977). In addition, we have found evidence for the de novo synthesis of trifoliin A in roots of axenically grown clover seedlings based on incorporation of labelled amino acids into a product which elutes from intact roots with 2-deoxyglucose, immunoprecipitates with anti-trifoliin A IgG, and comigrates with authenic trifoliin A in SDS-PAGE (Sherwood, Truchet, and Dazzo, manuscript in preparation).

Based on ultrastructural, biochemical, immunochemical, and genetic evidence, we present a revision of the original lectin cross-bridging model (Dazzo, Hubbell 1975b) on the specificity of initial attachment (Figure 1). In summary, trifoliin A reversibly cross-bridges receptors on the root hair cell wall with bacterial capsular polysaccharide and/or LPS. Trifoliin A and Truchet particles may literally play a central role in this initial attachment stage in that they are multivalent structures that can recognize and combine with unique saccharide sequences on both the microbe as well as the plant host. This revised model takes into account the growth phase-dependent LPS-trifoliin A interactions (Hrabak et al. 1981) which are predicted to be involved in infection (Kamberger 1979b), and the lectin involvement in the tip adhesion of adjacent root hairs of developing seedlings (Dazzo et al. 1982b; Higashi, Abe 1980).

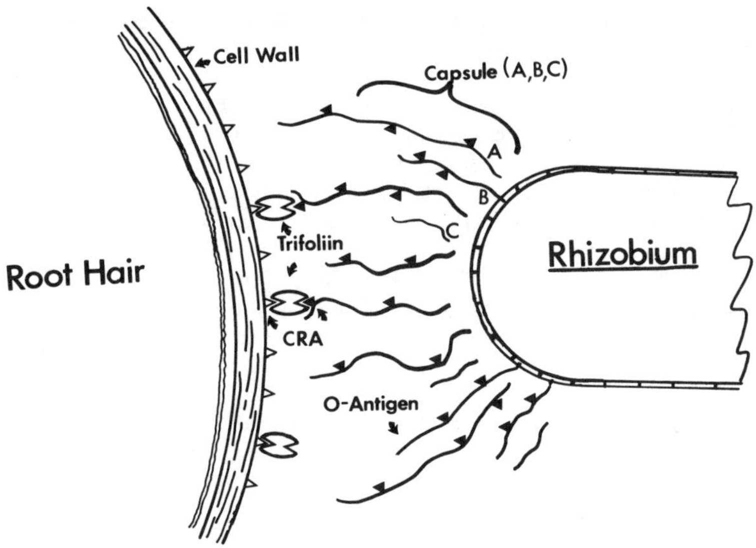

Figure 1 -- Lectin Cross-Bridging Model (Dazzo, Truchet 1983).

What is the evidence that the infection process at the root hair level is regulated? A common observation which indicates that root hair infection is tightly regulated is that only a few root hairs to which homologous rhizobia attach become infected. Thus, successful infection is a relatively rare event. This may be due to the transient infectibility of root hairs (Bhuvaneswari et al., 1981). Under controlled conditions in Fahraeus slide cultures, the number of root hairs infected by a strain of clover rhizobia will be characteristic of that strain, and may be different for other strains.

At least three components regulating lectin-Rhizobium interactions are known to exist in the Rhizobium-legume symbiosis. One is a plant component which is regulated by combined nitrogen, e.g., nitrate. A second is a bacterial component which is regulated by the growth phase (i.e., culture age) and has a profound effect on the ability of the bacterial polysaccharide to bind lectin. Finally, there is a less-well defined level of regulation which is detected only when the two symbionts are combined in situ.

Recognition of the rhizobial symbiont by the legume root is regulated by combined nitrogen (Dazzo, Brill 1978; Truchet, Dazzo 1982). The levels of trifoliin A on the root surface and attachment of R. trifolii to root hairs change dramatically when clover is grown with different levels of nitrate in the nutrient medium. During root growth at low levels of nitrate addition (1 mM), the levels of lectin and bacterial attachment increase when compared to seedlings grown with no combined N. These levels decline as nitrate is increased to approximately 15 mM. Plants growing at 15 mM are healthy. The nitrate effect on the root is not due to the counterion or to a direct interaction between nitrate and trifoliin A, and it requires more than 1 hr exposure to affect bacterial attachment (Dazzo, Hrabak 1982). Of potential importance to this effect, and the infection process in general, is the finding that clover and pea root cell walls change composition with nitrate supply (Dazzo et al. 1981; Diaz et al. 1981). Thus there must be some intervening mechanism which requires more than 1 hr to modulate this cell recognition process. Similar control of lectin levels by nitrate supply has been shown in the Azolla-Anabaena nitrogen-fixing symbiosis (Mellor et al. 1981).

The second example of regulation, which involves the culture age-dependent changes in polysaccharide composition, is under bacterial control. In R. japonicum 3I1b110, galactose residues in the capsular polysaccharide become methylated at carbon-4 during stationary phase, resulting in a reduction in soybean lectin binding (Mort, Bauer 1981). In R. trifolii 0403, at a specific time in early stationary phase, both the capsular polysaccharide and the LPS exhibit growth-phase dependent binding to trifoliin A which is 2-deoxyglucose reversible. The LPS undergoes profound immunochemical changes during normal batch culture growth (Hrabak et al. 1981). Unique antigenic determinants of the LPS which bind trifoliin A are transiently exposed on the bacterial surface just as the culture leaves lag phase, and again just as it enters stationary phase. Cells in exponential phase lack these LPS components and do not bind trifoliin A. Quantitative gas chromatography and combined GC-mass spectrometry showed several changes in the glycosyl composition of the LPS with culture age (Hrabak et al. 1981). Quinovosamine (2-amino-2,6-dideoxy-glucose) was one of the components confirmed to increase as the culture aged. Quinovosamine in the β-anomeric configuration was a major hapten of the unique LPS determinants which bound trifoliin A. Russa et al (1982) have shown that the clover Sym plasmid of R. trifolii controls incorporation of quinovosamine into its LPS. In addition, chemical changes in the capsular polysaccharide of R. trifolii 0403 have been found to occur with culture age (Sherwood, Abe, and Dazzo, manuscript in preparation). These growth-phase dependent chemical changes have an obvious impact on the validity of structural models of the repeating oligosaccharide units, and represent a major reason for inconsistent results from one laboratory to another on studies of bacterial polysaccharide-lectin interactions.

The final level of regulation occurs when the bacterium encounters the root environment of the host. For instance, some strains of R. japonicum bind soybean lectin in the root environment but not in pure culture (Bhuvaneswari, Bauer, 1978). We have recently found another level of regulation in the rhizosphere of clover while studying the curious delay in polar attachment which occurs when fully encapsulated cells of R. trifolii are used as inoculum (Dazzo et al. 1982a). While examining trifoliin A in clover root exudate of axenically grown plants (Dazzo, Hrabak 1981), we also discovered by serendipity an enzymatic activity in root exudate that alters and erodes the capsule in a way which favors polar attachment via polar-lectin receptors (Dazzo et al. 1982a). Because it had been suggested that lectins are enzymes which degrade the rhizobial polysaccharides (Albersheim, Wolpert, 1977), we fractionated clover root exudate by immunoaffinity chromatography to purify trifoliin A and assayed fractions for this enzymatic activity (Dazzo et al. 1982a). Reconstitution experiments showed that the enzymes responsible for this erosion of the capsule were antigenically unrelated to trifoliin A, and that trifoliin A did not fall into the special class of "enzymic lectins". Similarly, the N-acetylgalactosamine inhibitable soybean lectin (which specifically binds to the R. japonicum capsule) is not an enzymic lectin (Campello, Shannon 1982).

The process of capsule erosion was documented by

immunofluorescence and transmission electron microscopy (Dazzo et al. 1982a). This alteration begins in an equatorial band around the center of the bacterial cell and progresses out towards both poles at unequal rates, resulting in an intermediate condition where cells have capsular fibrils only at one pole. If the cells are heat-fixed to slides, all the capsule eventually erodes away. The kinetics of this enzymatic alteration are consistent with the time it takes for the transition of randomly oriented attachment to a predominantly polar attachment to clover root hairs. The equilibrium state of the nonadhering population of bacteria in the rhizosphere has trifoliin A bound in situ at one cell pole.

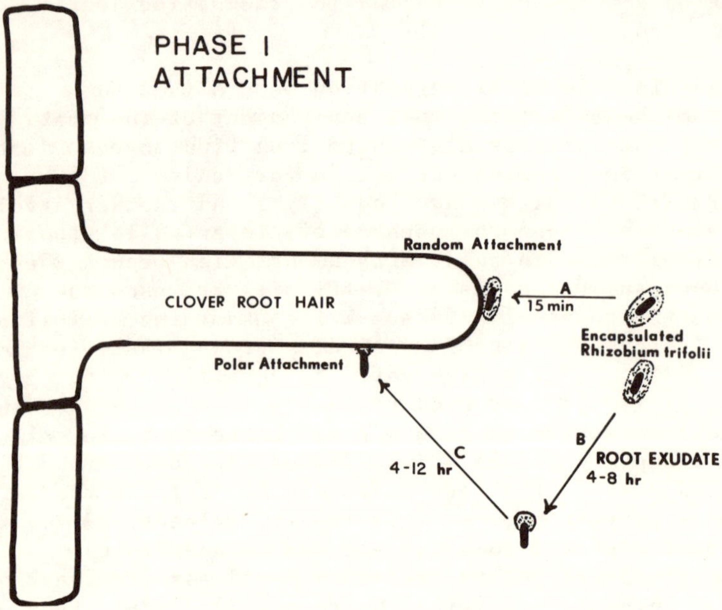

Figure 2 - Phase 1 Attachment (Dazzo, Truchet 1983)

From these data, we have constructed a dynamic, multi-phase model which is fully consistent with the timing and orientation of attachment of R. trifolii to clover root hairs (Figure 2). At high inoculum of fully encapsulated, lectin-binding cells, the bacteria attach within minutes to root hair tips where trifoliin A accumulates in greatest quantity (Phase 1A). During the next 4-8 hr, the capsule is altered by host enzymes in equilibrium with de novo bacterial polysaccharide synthesis, resulting in a cell with predominantly polar fibrils that bind trifoliin A (Phase 1B). We speculate that these cells with trifoliin A-binding receptors at one pole later polarly attach along the sides of the root hair, either to trifoliin A already bound to the root hair, or by binding to root exudate lectin and then to root hair lectin receptors (Phase 1C). Thus, polar attachment seems to be a manifestation of the location of the lectin receptor on the bacterium (Bohlool, Schmidt 1976; Truchet, Sherwood, Dazzo & Vasse, manuscript in preparation).

The above discussion directly supports the validity of the lectin recognition hypothesis as originally proposed by Bohlool & Schmidt (1974). However, the most important work which has countered this hypothesis is the finding that soybean lines which lack seed lectin nodulate normally with R. japonicum (Pull et al. 1979). This latter discrepancy may be resolved by the recent detection of small quantities of lectin in seed and of lectin messenger RNA in developing seedling roots of these same soybean lines (Schmidt 1979; Goldberg et al., in preparation). The reader is referred to the reviews of Bauer (1980), Schmidt (1979), Dazzo (1981), and Dazzo & Truchet (1983) for lengthy discussions on other pros and cons of the lectin recognition hypothesis.

In summary, some legume lectins fulfill many criteria that are expected of a cell recognition molecule involved in the nitrogen-fixing Rhizobium-legume symbiosis. However, the research accomplishments of the last decade testing the lectin-recognition hypothesis are only beginning to answer questions on the infection process. It is clear that lectin-mediated bacterial attachment is only one of the steps necessary for successful infection, since root hair infection can also be genetically blocked at other locations. The lectin recognition hypothesis will never be universally accepted or rejected until the appropriate bacterial and plant mutants are available to focus directly on the role of the Rhizobium-binding lectin and its saccharide receptors in the recognition code without affecting other events in the infection process.

ACKNOWLEDGMENTS

Work in the author's lab was supported by NSF grant 80-21906 and USDA-CRGO Grant 82-CRCR-1-1040. This is Michigan Agricultural Experiment Station Journal No.

REFERENCES

Albersheim P, Wolpert J (1977) Molecular determinants of symbiont-host selectivity between nitrogen fixing bacteria and plants. In Solheim B and Raa J (eds): "Cell Wall Biochemisry Related to Specificity in Host-Plant Pathogen Interactions", Tromso: Universiteitsforlanget, pp. 373-375.

Bauer WD (1981) Infection of legumes by rhizobia. Ann. Rev. Plant Physiol 32:407.

Bhuvaneswari TV, Bauer WD (1978) The role of lectins in plant-microorganism interactions. III. Influence of rhizosphere/rhizoplane culture conditions on the soybean lectin-binding properties of rhizobia. Plant Physiol 62:71.

Bhuvaneswari TV, Bhagwat AA, Bauer WD (1981). Transient susceptibility of root cells in four common legumes to nodulation by rhizobia. Plant Physiol 68:1144.

Bhuvaneswari TV, Pueppke SG, Bauer WD (1977) Role of lectins in plant-microorganism interactions. I. Binding of soybean lectin to rhizobia. Plant Physiol 60:486.

Bishop PE, Dazzo FB, Applebaum ER, Maier RJ, Brill WJ (1977) Intergeneric transformation of genes involved in the Rhizobium-legume symbiosis. Science 198:938.

Campello ED, Shannon LM (1982) An α-galactosidase with hem-
agglutinin properties from soybean seeds. Plant Physiol 69,628.

Bohlool BB, Schmidt, EL (1974) Lectins: a possible basis for specificity in the Rhizobium-legume root nodule symbiosis. Science 185:269.

Bohlool BB, Schmidt EL (1976) Immunofluorescent polar tips of Rhizobium japonicum: possible site of attachment or lectin binding. J Bacteriol 125:1188.

Chen AP, Phillips DA (1976) Attachment of Rhizobium to legume roots as the basis for specific associations. Physiol Plant 38:83.

Dazzo FB (1981) Bacterial attachment as related to

cellular recognition in the Rhizobium-legume symbiosis. J Supramol Struct Cell Biochem 16:29.

Dazzo FB, Brill WJ (1977) Receptor site on clover and alfalfa roots for Rhizobium. Appl Environ Microbiol 33:132.

Dazzo FB, Brill WJ (1978) Regulation by fixed nitrogen of host-symbiont recognition in the Rhizoibum-clover symbiosis. Plant Physiol 62:18.

Dazzo FB, Brill WJ (1979) Bacterial polysaccharide which binds Rhizobium trifolii to clover root hairs. J Bacteriol 137:1362.

Dazzo FB, Hrabak EM (1981) Presence of trifoliin A, a Rhizobium-binding lectin, in clover root exudate. J Supramol Struct Cell Biochem 16:133.

Dazzo FB, Hrabak EM (1982) Lack of a direct interaction between nitrate and trifoliin A in the Rhizobium-clover symbosis. Plant Soil 69,259.

Dazzo FB, Hrabak EM, Urbano MR, Sherwood JE, Truchet GL (1981) Regulation of recognition in the Rhizobium-clover symbiosis. In Gibson AH, Newton WE (eds): "Current Perspectives in Nitrogen Fixation", Canberra: Australian Academy of Science, pp. 292-295.

Dazzo FB, Hubbell DH (1975a) Concanavalin A: lack of correlation between binding to Rhizobium and host specificity in the Rhizobium-legume symbiosis. Plant Soil 43,713.

Dazzo FB, Hubbell DH (1975b) Cross-reactive antigens and lectin as determinants of symbiotic specificity in the Rhizobium-clover association. Appl Microbiol 30:1017.

Dazzo FB, Napoli CA, Hubbell DH (1976) Adsorption of bacteria to roots as related to host specificity in the Rhizobium-clover association. Appl Environ Microbiol 32:168.

Dazzo FB, Truchet GL (1983) Interactions of lectins and their saccharide receptors in the Rhizobium-legume symbiosis. J Memb Biol 73:1

Dazzo FB, Truchet GL, Kijne JW (1982) Lectin involvement in root hair tip adhesions as related to the Rhizobium-clover symbiosis. Physiol Plant 56,143

Dazzo FB, Truchet GL, Sherwood JE, Hrabak EM, Gardiol AE (1982) Alteration of the trifoliin A-binding capsule of Rhizobium trifolii 0403 by enzymes released from clover roots. Appl Environ Microbiol 44:478.

Dazzo FB, Urbano MR, Brill WJ (1979) Transient appearance of lectin receptors on Rhizobium trifolii. Curr Microbiol. 2:15.

Dazzo FB, Yanke WE, Brill WJ (1978) Trifoliin: a Rhizobium recognition protein from white clover. Biochim Biophys Acta 536:276.

Diaz C, Kijne JW, Quispel A (1981) Influence of nitrate on pea root cell wall composition. In Gibson AH, Newton WE (eds): "Current Perspectives in Nitrogen Fixation", Canberra: Australian Academy of Science, p. 426.

Gade W, Jack MA, Dahl JB, Schmidt EL, Wolf F (1981) The isolation and characterization of a root lectin from soybean (Glycine max L.) cultivar Chippewa. J Biol Chem 256:12905.

Hamblin J, Kent SP (1974) Possible role of phytohaemagglutinin in Phaseolus vulgaris L. Nature New Biol 245:28.

Higashi S, Abe M (1980) Scanning electron microscopy of Rhizobium trifolii infection sites on root hairs of white clover. Appl Environ Microbiol 40:1094.

Hooykaas PJ, van Brussell AAN, den Hulk-Ras H, van Slogteren GM, Schilperoort RA (1981) Sym plasmid of Rhizobium trifolii expressed in different rhizobia species and Agrobacterium tumefaciens. Nature 291:351.

Hrabak EM, Urbano MR, Dazzo FB (1981) Growth-phase dependent immunodeterminants of Rhizobium trifolii lipopolysaccharide which bind trifoliin A, a white clover lectin. J Bacteriol 148:697.

Hughes TA, Elkan GH (1981) Study of the Rhizobium japonicum-soybean symbiosis. Plant Soil 61:87.

Johnston AWB, Beyon IL, Buchanan-Wollaston AV, Setchell SM, Hirsch PR, Beringer JE (1978) High frequency transfer of nodulating ability between species and strains of Rhizobium. Nature 276:635.

Kamberger W. (1979a) An Ouchterlony double diffusion study on the interaction between legume lectins and rhizobial cell surface antigens. Arch Microbiol 121:83.

Kamberger W. (1979b) Role of cell surface polysaccharides in the Rhizobium-pea symbiosis. FEMS Microbiol Lett 6:361

Kato G, Maruyama Y, Nakamura M (1980) Role of bacterial polysaccharides in the adsorption process of the Rhizobium-pea symbiosis. Agric Biol Chem 44:2843.

Kato G, Maruyama Y, Nakamura M (1981) Involvement of lectins in Rhizobium-pea recognition. Plant Cell Physiol 22:759.

Kijne JW, van der Schaal IAM, de Vries GE (1980) Pea lectins and the recognition of Rhizobium leguminosarum. Plant Sci Lett 18:65.

Li D, Hubbell DH (1969) Infection thread formation as a basis for nodulation specificity in Rhizobium-strawberry clover associations. Can J Microbiol 15,1133.
Mellor RB, Gadd GM, Rowell P, Stewart WDP (1981) A phytohemagglutinin from the Azolla-Anabaena symbiosis. Biochem Biophys Res Commun 99,1348.
Mort AJ, Bauer WD (1980) Composition of the capsular and extracellular polysaccharides of Rhizobium japonicum: changes with culture age and correlations with binding of soybean seed lectin to the bacteria. Plant Physiol 66:158
Paau AS, Leps WT, Brill WJ (1981) Agglutinin from alfalfa necessary for binding and nodulation by Rhizobium meliloti. Science 213:1513.
Pueppke SG, Freund TG, Schutly BC, Friedman HP (1980) Interaction of lectins from soybean and peanut with rhizobia that nodulate soybean, peanut, or both plants. Can J Microbiol 26,1489.
Pull SP, Pueppke SG, Hymowitz T, Orf JH (1978) Soybean lines lacking the 120,000 dalton seed lectin. Science 200:1277.
Russa R, Urbanik T, Kowalczuk E, Lorkiewicz Z (1982) Correlation between occurrence of plasmid pUCS202 and lipopolysaccharide alterations in Rhizobium. FEMS Microbiol Lett 13:161.
Schmidt EL (1979) Initiation of plant root-microbe interactions. Ann Rev Microbiol 33:335.
Stacey G, Paau AS, Brill WJ (1980) Host recognition in the Rhizobium-soybean symbiosis. Plant Physiol 66:609.
Truchet GL, Dazzo FB (1982) Morphogenesis of lucerne root nodules incited by Rhizobium meliloti in the presence of combined nitrogen. Planta 154:352.
van Rensberg HJ, Strijdom BW (1982) Root surface association in relation to nodulation of Medicago sativa. Appl Environ Microbiol 44:93.
Wolpert JS, Albersheim P (1976) Host-symbiont interactions. I. The lectins of legumes interact with the O-antigen-containing lipopolysaccharides of their symbiont rhizobia. Biochem Biophys Res Commun 70:729.
Zurkowski W (1980) Specific adsorption of bacteria to clover root hairs, related to the presence of the plasmid pWZ2 in cells of Rhizobium trifolii. Microbios 27:27.

Chemical Taxonomy, Molecular Biology, and Function
of Plant Lectins, pages 225–236

SOYBEAN LECTIN: DOES IT HAVE AN ESSENTIAL ROLE IN THE RHIZOBIUM-SOYBEAN SYMBIOSIS?

Steven G. Pueppke

Department of Plant Pathology
University of Florida
Gainesville, FL 32611

Soybean lectin (SBL) is a well-characterized glycoprotein lectin from seeds of the cultivated soybean, *Glycine max* (L.) Merr. The lectin is a tetramer, and it displays hapten specificity for N-Acetyl-D-galactosamine (galNAc) and to a lesser extent for D-galactose (gal) (Pereira et al. 1974). Although the native lectin and several labeled derivatives are used extensively as probes of carbohydrate structure, the activity of SBL in the plant remains uncertain. The concentration of SBL in the seed can be quite substantial (Liener 1976; Pueppke et al. 1978; Pull et al. 1978), and it stands to reason that the *in situ* role of the seed lectin is related in some way to its ability to bind specific saccharides. Likely functions of seed SBL thus include (i) maintenance of cell wall or membrane integrity, (ii) transport or storage of gal- or galNAc-containing molecules, (iii) roles in seed development or germination, and (iv) defense against fungi, bacteria, or insects that attack soybean seeds (see Goldstein 1981; Sharon, Lis 1972). About 10 years ago Bohlool and Schmidt (1974) hypothesized that SBL plays an indispensible role in the inception of the nitrogen-fixing soybean-*Rhizobium* symbiosis. The "lectin recognition hypothesis" of Bohlool and Schmidt commanded immediate attention, and even though seed SBL is used in all experiments with rhizobia, emphasis was shifted from the seed to the roots. As a consequence the study of seed SBL lost momentum, and research on SBL relative to the symbiosis began to outpace all other approaches. It is my objective here to evaluate the lectin recognition hypothesis in light of data from the literature and some recent experiments from my laboratory.

The soybean-*Rhizobium* symbiosis is initiated when compatible *Rhizobium* cells associate with the surfaces of soybean root hairs. The rhizobia enter by means of infection threads--tubular structures encasing *Rhizobium* cells like links of a sausage. The infection threads elongate toward the interior of the root, and the rhizobia eventually are released into developing root nodules (Dart 1974). Soybean root hairs accept cells of only certain *Rhizobium* strains. All strains of *R. japonicum* ordinarily infect soybean, as do some strains classified as *R. lupini* or relegated to an ill-defined group termed *Rhizobium* spp. (Trinick 1982; Vincent 1977). Other rhizobia do not nodulate soybean, and they do not initiate infection threads in soybean root hairs (Pueppke 1983). The cellular and biochemical transactions that allow soybean to differentiate infective from noninfective rhizobia thus take place on or near the root surface. The hypothesis first advanced by Bohlool and Schmidt (1974) and later elaborated by Schmidt (1979) and others (Stacey, Brill 1982; Stacey et al. 1980) states that soybean root surfaces contain SBL and that the lectin reacts in a hapten-specific manner with saccharides on the surfaces of nodulating (but not nonnodulating) *Rhizobium* cells. Nodulating *Rhizobium* cells are proposed to attach to soybean root surfaces by means of such complementary lectin-carbohydrate binding, and the resulting cell-cell contact is thought to facilitate infection thread formation.

Two presuppositions are implicit in the lectin recognition hypothesis, and evidence bearing on each has been used to evaluate the hypothesis. The first assumption is quite simple: if SBL accounts for the specificity of the symbiosis, then the lectin must be present in soybean roots competent to be infected by rhizobia. The second assumption is that all *Rhizobium* strains capable of infecting soybean ought to bind SBL. Although Bhuvaneswari (1980) argues that SBL also might bind fortuitously to some *Rhizobium* strains incapable of infecting soybean, Bohlool and Schmidt (1974) and Stacey et al. (1980) both emphasize the lack of interaction between the lectin and soybean nonsymbionts.

BINDING OF SBL TO RHIZOBIA

Three methods commonly are used to assess the binding of SBL to *Rhizobium* cells. In the first, SBL is labeled with a fluorochrome such as fluorescein isothiocyanate. Suitable dilutions of the fluorescent-labeled lectin then are mixed with cells and binding is evaluated with a fluorescence microscope. This procedure is suitable for screening the SBL-binding capacities of large numbers of strains and in theory can detect about 2,000 SBL molecules bound per bacterium (Bhuvaneswari, Bauer 1978). Fluorescence microscopy also yields information about percentages of lectin-binding cells in the bacterial population and about the location of lectin receptors on the Rhizobium cell surface (Bhuvaneswari, Bauer 1978; Bhuvaneswari et al. 1977; Bohlool, Schmidt 1974; Bohlool, Schmidt 1976; Chen, Phillips 1976; Law et al 1982; Mort, Bauer 1980; Pueppke et al. 1980). Although it is possible to use fluorescence intensities to estimate the extent of binding, interpretation is subjective, and the data are semiquantitative at best. SBL can be labeled with electron-dense metals and visualized with the electron microscope (Bal et al. 1978; Calvert et al. 1978; Shantharam et al. 1980; Tsien, Schmidt 1981). The data allow SBL receptors to be pinpointed with high resolution, but the method is inappropriate for large scale surveys. Radioisotope-labeled SBL also has been prepared and used to measure lectin binding (Bhuvaneswari et al. 1977; Brethauer, Paxton 1977; Meyer, Pueppke 1980; Pueppke et al. 1980). The method is quantitative and reproducible and can be used to monitor temporal fluctuations in the ability of rhizobia to bind the lectin.

SBL adsorbs nonspecifically to plastic (Bhuvaneswari et al. 1977; Pueppke et al. 1980; Su et al. 1980) and also apparently to some bacterial surfaces (Bhuvaneswari, Bauer 1978). It thus is imperative that controls for hapten specificity be included in binding studies, regardless of the lectin labeling method used. As a matter of course, the lectin also should be purified to homogeneity. Unfortunately, rigorous criteria for SBL purity and binding specificity have not been established in some assays of SBL binding to rhizobia. In several early screening studies (Bohlool, Schmidt 1974; Law, Strijdom 1977), for example, fluorescent-labeled "SBL" consisted of a crude ammonium sulfate precipitate of seed proteins, which has been estimated to be 85% contaminated by extraneous proteins (Brethauer, Paxton 1977). I believe that these and similar studies have led to several misconceptions about SBL binding to rhizobia. Table 1 is a compilation of binding

Table 1
Binding of SBL to Rhizobium cells[a]

Species	Number of strains which are: Positive	Negative	Inconsistent[b]
R. japonicum	36	39	3
R. lupini	1	9	0
R. spp.	4	31	2
R. leguminosarum	4	8	1
R. meliloti	0	18	1
R. phaseoli	5	7	1
R. trifolii	5	7	0
TOTALS	55	119	8

[a] The data are compiled from Bhuvaneswari, Bauer (1978); Bhuvaneswari et al. (1977); Bohlool, Schmidt (1974); Brethauer, Paxton (1977); Chen, Phillips (1976); Kato et al. (1979); Law, Strijdom (1977); and Pueppke et al. (1980). Included are the results of recent screening experiments performed in my laboratory as described by Pueppke et al. (1980). In these experiments, R. japonicum 73, 91, 136, and RJ17W bound SBL. The following strains did not bind SBL: R. japonicum RJ4G, RJ7AS, RJ14C, 23, 25, 33, 61, 84, 94, 100, 119, 191, 192, 193, 194, 201, 205, 206, 208, 214, 217; Rhizobium spp. 94A3, 94A4, 94A6, 94A7, 176B13, 229, 3263; R. meliloti 102F26, 102F29, 102F51, 102F65, 1023, 1045; R. phaseoli 127K14; R. trifolii 0403.

[b] Different laboratories report that these strains both bind and fail to bind SBL.

data from the literature, including results of experiments where hapten specificity was not established (Bohlool, Schmidt 1974; Chen, Phillips 1976; Kamberger 1979; Kato et al. 1979; Law, Strijdom 1977), where SBL was impure (Bohlool, Schmidt 1974; Law, Strijdom 1977), or where source of the bacterial lectin receptor was ambiguous (Bhuvaneswari, Bauer 1978; Graham 1981). I have added results of recent experiments from my laboratory, which were performed as described by Pueppke et al. (1980) using fluorescent-labeled SBL. Although 182 Rhizobium strains are listed in Table 1, the vast majority are either R. japonicum or Rhizobium spp. Members of all Rhizobium species except R. meliloti bind the "lectin," and about 48% of symbiotic R.

Table 2
Hapten-specific Binding of Affinity-purified SBL to *Rhizobium* cells[a]

Species	Number of strains which are: Positive	Negative	Inconsistent[b]
R. japonicum	24	32	2
R. lupini	0	1	0
R. spp.	4	28	1
R. leguminosarum	0	0	0
R. meliloti	0	7	0
R. phaseoli	0	3	0
R. trifolii	0	2	0
TOTALS	28	73	3

[a] The data are compiled from Bhuvaneswari et al. (1977); Brethauer, Paxton (1977); and Pueppke et al. (1980). Recent results from my laboratory are included (see footnotes to Table 1).

[b] Different laboratories report that these strains both bind and fail to bind SBL.

japonicum and 42% of nonsymbiotic *R. phaseoli* and *R. trifolii* are lectin-positive. In addition, 8 strains are reported both to bind and not to bind the lectin.

When the questionable data from Table 1 are excluded, the picture is much more clearcut (Table 2). Most of the laboratory to laboratory inconsistencies are removed, as are *all* of the reports that nonnodulating *R. lupini*, *R. leguminosarum*, *R. phaseoli*, and *R. trifolii* bind the lectin. The percentages of *R. japonicum* and *Rhizobium* spp. strains capable of binding SBL also are reduced somewhat. This leads me to suggest that a substantial portion of the SBL binding reported in Table 1 is nonspecific or represents the adsorption of labeled contaminating proteins to the bacterial cells. The majority of symbiotic *R. japonicum* strains nevertheless fail to bind affinity-purified SBL in a hapten-specific manner, and of the 4 *Rhizobium* spp. strains that bind SBL, only one (strain 3G4b16) nodulates soybean (Brethauer, Paxton 1977; Pueppke et al 1980).

The competence of rhizobia to bind SBL is strongly dependent on culture age and growth medium (Bhuvaneswari,

Bauer 1978; Bhuvaneswari et al. 1977, Pueppke et al. 1980). Manipulation of growth conditions is known to convert some of the binding strains in Table 2 to the nonbinding condition (Bhuvaneswari et al. 1977; Pueppke et al. 1980), and it is reasonable that the reverse could occur, i.e. that some nonbinding strains could be rendered competent to bind the lectin. Thirty-three strains of Rhizobium from our culture collection, including 15 slow-growing R. japonicum strains, 10 fast-growing R. japonicum strains from the People's Republic of China, and 6 R. meliloti strains, nevertheless have resisted all of our efforts (modification of growth medium, sampling at various culture ages) to coax them to bind SBL.

DO SOYBEAN ROOTS CONTAIN SBL?

There are several reports that crude extracts of soybean roots contain hemagglutinating activity (Bohlool, Schmidt 1974; Bowles et al. 1979; Kato et al. 1979; Pueppke et al. 1978), but such activity is insufficient reason to conclude that the roots contain SBL. Plants are known to produce both nonprotein hemagglutinins and proteins that agglutinate in a saccharide-independent manner (Bowles et al. 1979; Callow 1975; Fett and Sequeira 1980; Leach et al. 1982), and neither substance properly can be termed a lectin (Goldstein et al. 1980). In addition, the detergents sometimes used to solubilize agglutinating activity from soybean roots (Bowles et al. 1979) are recognized to interfere with hemagglutination assays (Lotan et al. 1977; Pueppke et al. 1981). Although Pueppke et al. (1978) corroborated their initial hemagglutination data by radioimmunoassay, Keegstra (personal communication) and Su et al. (1980) were unable to reproduce the results by analyzing new soybean root extracts. We now believe that the original root extracts (Pueppke et al. 1978) were contaminated by SBL from an external source.

Gade et al. (1981) attempted to preparatively isolate and characterize SBL from Chippewa soybean roots. Seeds were germinated for 5 days on moist cheesecloth, roots were excised and rinsed, and proteins were eluted with a buffer containing gal. The hapten solubilized approximately 0.5% of the dry weight of the roots as protein, and SBL was purified from this extract by sequential affinity chromatography on immobilized galNAc and Concanavalin A. Recovery of SBL was about 50 ng/g fresh weight of root tissue. The galNAc column purified two hemagglutinins--the activity of the first was reduced (but evidently not abolished) by galNAc and gal, and the activity of the second was insensitive to the haptens. Unfortunately, the data of Gade et al. (1981) are inconclusive, because the conditions used for germination of the seeds are known to promote leaching of SBL from seeds and cotyledons (Fountain et al. 1977; Hwang et al. 1978). The lectin associated with the roots thus simply may have been a seed lectin contaminant (the amount of SBL in the cotyledons of one 5-day-old soybean seedling is about 12 times that recovered from 200 g of roots). Gade et al. (1981) attempted to segregate roots from cotyledons in two control experiments, and hemagglutinating activity was detected in extracts of such roots. Hapten controls were omitted from these assays, and thus the second contaminating agglutinin may have been responsible for the hemagglutinating activity.

Although SBL was detected immunologically on root hairs of Corsoy soybean seedlings (Stacey et al. 1980), radioimmunoassays failed to detect the lectin in carefully prepared buffer extracts of 5-day-old roots of 20 soybean varieties and 3- to 66-day-old roots of 2 varieties (Su et al. 1980). Bauer (1980) and Bhuvaneswari (1980) similarly could not find SBL on the surfaces of soybean roots. Such roots nevertheless are readily infected and nodulated by soybean rhizobia (Pueppke 1983; Su et al. 1980). Several of the soybean lines that lack detectable root lectin also are devoid of seed SBL (Pueppke et al. 1981; Pull et al. 1978; Stacey et al. 1981; Su et al. 1980). These lines are homozygous recessive at the locus designated *le* (Orf et al. 1978), and the structural gene for SBL contains a 3.4 kilobase insertion that blocks transcription (Goldberg et al. 1983; Vodkin 1981).

ADHESION OF RHIZOBIA TO SOYBEAN ROOTS

If symbiotic rhizobia anchor to soybean roots by means of SBL, addition of haptens or soluble SBL would be expected to disrupt bacterial attachment. Only a few such bacterial adsorption experiments have been done, and the results as yet are inconclusive. Stacey et al. (1980) found that 60 mM gal or 30 mM galNAc abolished the binding of *R. japonicum* strain 110 to elongated root hairs of the wild soybean, *Glycine soja*. Bauer (1980, 1982), however, observed no consistent effect of haptens on the attachment of rhizobia to soybean roots, and he suggested that exogenously supplied galNAc actually may enhance nodulation. We recently described a method to measure the binding of rhizobia to soybean roots (Pueppke, Benny 1983). The assay has been used to monitor the influence of exogenously supplied, soluble SBL on attachment of rhizobia to soybean roots. The inoculum contains 10^4 Rhizobium cells and 10 µg of SBL per ml. The SBL:Rhizobium cell ratio thus is about 5×10^9 molecules/cell, and the lectin receptors on the bacterial cells theoretically are saturated with the lectin (Bhuvaneswari et al. 1977). Two Rhizobium strains were examined. R. lupini 96B9 neither infects soybean nor binds SBL (Bhuvaneswari et al. 1977; Pueppke 1983), and the binding rate of strain 96B9 to soybean roots (155 bacteria bound per 2-cm root segment/hour) is unchanged by the presence of the lectin. In contrast, R. japonicum 138 infects soybean root hairs and binds SBL, and the lectin decreases the binding rate of strain 138 from 35 to 22 bacteria bound per 2-cm root segment/hour. The magnitude of the binding reduction, however, is not great. We also do not yet know if the effect of SBL on attachment of 138 is specific or simply reflects perturbation of the bacterial surface, e.g. changes in charge distribution or the hydrophobic character of the bacterial lipopolysaccharides or capsules. It also is noteworthy that soybean roots bind greater numbers of noninfective strain 96B9 than of infective strain 138!

CONCLUSIONS

The lectin recognition hypothesis states that "lectins present a site on the legume root surface that interacts specifically with a distinctive saccharide at the surface of the appropriate Rhizobium cell as a prelude to nodulation" (Schmidt 1979). That SBL serves as such a recognition site is disputed by four observations:

1. The binding of SBL to soybean-nodulating rhizobia is merely selective, and not specific.
2. Many Rhizobium strains nodulate soybean roots but do not bind detectable amounts of SBL.
3. Soybean roots that lack detectable SBL are competent to form infection threads and nodules.
4. Exogenously-supplied SBL or haptens of SBL have rather unimpressive effects on the adsorption of rhizobia to

soybean roots.
Each observation is of course provisional and subject to reinterpretation in the light of additional data. The available evidence nevertheless fails to provide compelling reason to accept the hypothesis that SBL plays a specific and indispensible role in the soybean-*Rhizobium* symbiosis.

REFERENCES

Bal AK, Shantharam S, Ratnam S (1978). Ultrastructure of *Rhizobium japonicum* in relation to its attachment to root hairs. J Bacteriol 133:1393.

Bauer WD (1980). Role of soybean lectin in the soybean-*Rhizobium japonicum* symbiosis. In Newton WJ, Orme-Johnson, WH (eds): "Nitrogen Fixation, Volume II," Baltimore: University Park Press, p 205.

Bauer WD (1982). Attachment of rhizobia to soybean roots. Plant Physiol 69:Suppl 143 (Abstract).

Bhuvaneswari TV (1980). Recognition mechanisms and infection process in legumes. Econ Bot. 35:204.

Bhuvaneswari TV, Bauer WD (1978). Role of lectins in plant-microorganism interactions III. Influence of rhizosphere/rhizoplane culture conditions on the soybean lectin-binding properties of rhizobia. Plant Physiol 62:71.

Bhuvaneswari TV, Pueppke SG, Bauer WD (1977). Role of lectins in plant-microorganism interactions I. Binding of soybean lectin to rhizobia. Plant Physiol 60:486.

Bohlool BB, Schmidt EL (1974). Lectins: a possible basis for specificity in the *Rhizobium*-legume root nodule symbiosis. Science 185:269.

Bohlool BB, Schmidt EL (1976). Immunofluorescent polar tips of *Rhizobium japonicum*: possible site of attachment or lectin binding. J Bacteriol 125:1188.

Bowles DJ, Lis H, Sharon N (1979). Distribution of lectins in membranes of soybean and peanut plants I. General distribution in root, shoot and leaf tissue at different stages of growth. Planta 145:193.

Brethauer TS, Paxton JD (1977). The role of lectin in soybean-*Rhizobium japonicum* interactions. In Solheim B, Raa J (eds): "Cell wall biochemistry related to specificity in host-plant pathogen interactions," Oslo: Universitetsforlaget, p 381.

Callow JA (1975). Plant lectins. Curr Adv Plant Sci 18:221.

Calvert, HE, Lalonde M, Bhuvaneswari TV, Bauer WD (1978). Role of lectins in plant-microorganism interactions IV. Ultrastructural localization of soybean lectin binding sites on Rhizobium japonicum. Can J Microbiol 24:785.

Chen APT, Phillips DA (1976). Attachment of Rhizobium to legume roots as the basis for specific interactions. Physiol Plant 38:83.

Dart PJ (1974). Development of the root-nodule symbiosis. In Quispel A (ed): "The biology of nitrogen fixation," Amsterdam: North-Holland Publishing Co, p 381.

Fett WF, Sequeira L (1980). A new bacterial agglutinin from soybean I. Isolation, partial purification, and characterization. Plant Physiol 66:847.

Fountain DW, Foard DE, Replogle WD, Yang WK (1977). Lectin release by soybean seeds. Science 197:1185.

Gade W, Jack MA, Dahl JB, Schmidt EL, Wold F (1981). The isolation and characterization of a root lectin from soybean (Glycine max (L), cultivar Chippewa). J Biol Chem 256:12905.

Goldberg RB, Jofuku D, Vodkin L (1983). Seed protein gene structure in normal and mutant soybean lines. In Ahmad F, Downey K, Schultz J, Voellmy RW (eds): "Advances in gene technology: molecular genetics of plants and animals," Miami: University of Miami, p 30 (Abstract).

Goldstein IJ (1981). Plant derived lectins. Rec Adv Phytochem 15:25.

Goldstein IJ, Hughes RC, Monsigny M, Osawa T, Sharon N (1980). What should be called a lectin? Science 285:66.

Graham TL (1981). Recognition in Rhizobium-legume symbioses. Int Rev Cytol 13:127.

Hwang DL, Yang WK, Foard DE, Lin KTD (1978). Rapid release of protease inhibitors from soybeans. Immunochemical quantitation and parallels with lectins. Plant Physiol 61:30.

Kamberger W (1979). An Ouchterlony double diffusion study on the interaction between legume lectins and rhizobial cell surface antigens. Arch Microbial 121:83.

Kato G, Maruyama Y, Nakamura M (1979). Role of lectins and lipopolysaccharides in the recognition process of specific legume-Rhizobium symbiosis. Agric Biol Chem 43:1085.

Law IJ, Strijdom BW (1977). Some observations on plant lectins and Rhizobium specificity. Soil Biol Biochem 9:79.

Law IJ, Yamamoto Y, Mort AJ, Bauer WD (1982). Nodulation of soybean by Rhizobium japonicum mutants with altered

capsule synthesis. Planta 154:100.
Leach JE, Cantrell MA, Sequeira L (1982). Hydroxyproline-rich bacterial agglutinin from potato. Extraction, purification, and characterization. Plant Physiol 70:1353.
Liener IE (1976). Phytohemagglutinins (phytolectins). Annu Rev Plant Physiol 27:291.
Lotan R, Beattie G, Hubbell W, Nicolson GL (1977). Activities of lectins and their immobilized derivatives in detergent solutions. Implications on the use of lectin affinity chromatography for the purification of membrane glycoproteins. Biochemistry 16:1787.
Meyer MC, Pueppke SG (1980). Differentiation of Rhizobium japonicum strain derivatives by antibiotic sensitivity patterns, lectin binding, and utilization of biochemicals. Can J Microbiol 26:606.
Mort AJ, Bauer WD (1980). Composition of the capsular and extracellular polysaccharides of Rhizobium japonicum. Changes with culture age and correlations with binding of soybean seed lectin to bacteria. Plant Physiol 66:158.
Orf JH, Hymowitz T, Pull SP, Pueppke SG (1978). Inheritance of a soybean seed lectin. Crop Sci 18:899.
Pereira MEA, Kabat EA, Sharon N (1974). Immunochemical studies on the specificity of soybean agglutinin. Carbohydr Res 37:89.
Pueppke SG (1983). Rhizobium infection threads in root hairs of Glycine max (L.)Merr., Glycine soja Sieb. & Zucc., and Vigna unguiculata (L.)Walp. Can J Microbiol 29:69.
Pueppke SG, Bauer WD, Keegstra K, Ferguson AL (1978). Role of lectins in plant-microorganism interactions II. Distribution of soybean lectin in tissues of Glycine max (L.)Merr. Plant Physiol 61:779.
Pueppke SG, Benny UK (1983). Adsorption of infective and noninfective rhizobia to infectible portions of soybean roots. Plant Physiol 71: (abstract in press).
Pueppke SG, Freund TG, Schulz BC, Friedman HP (1980). Interaction of lectins from soybean and peanut with rhizobia that nodulate soybean, peanut, or both plants. Can J Microbiol 26:1489.
Pueppke SG, Friedman HP, Su LC (1981). Examination of Le and lele genotypes of Glycine max (L.)Merr. for membrane-bound and buffer-soluble soybean lectin. Plant Physiol 68:905.
Pull SP, Pueppke SG, Hymowitz T, Orf JH (1978). Soybean lines lacking the 120,000-dalton seed lectin. Science

200:1277.
Schmidt EL (1979). Initiation of plant root-microbe interactions. Annu Rev Microbiol 33:355.
Shantharam S, Gow JA, Bal AK (1980). Fractionation and characterization of two morphologically distinct types of cells in Rhizobium japonicum broth culture. Can J Microbiol 26:107.
Sharon N, Lis H (1972). Lectins: cell-agglutinating and sugar-specific proteins. Science 177:949.
Stacey G, Brill WJ (1982). Nitrogen-fixing bacteria: colonization of the rhizosphere and roots. In Mount MS, Lacy GH (eds): "Phytopathogenic prokaryotes, Volume 1," New York: Academic Press, p 225.
Stacey, G, Paau AS, Brill WJ (1980). Host recognition in the Rhizobium-soybean symbiosis. Plant Physiol 66:609.
Stacey G, Paau AS, Brill WJ (1981). Host recognition in the Rhizobium-soybean symbiosis. Plant Physiol 67:Suppl 134 (Abstract).
Su LC, Pueppke SG, Friedman HP (1980). Lectins and the soybean-Rhizobium symbiosis I. Immunological investigations of soybean lines, the seeds of which have been reported to lack the 120,000 dalton soybean lectin. Biochim Biophys Acta 629:292.
Trinick MJ (1982). Host-Rhizobium associations. In Vincent JM (ed): "Nitrogen fixation in legumes," Sydney: Academic Press, p 111.
Tsien HC, Schmidt EL (1981). Localization and partial characterization of soybean lectin-binding polysaccharide of Rhizobium japonicum. J Bacteriol 145:1063.
Vincent JM (1977). Rhizobium: general microbiology. In Hardy RFW, Silver, WS (eds): "A treatise on dinitrogen fixation. Section III: biology," New York: Wiley-Interscience, p 277.
Vodkin LO (1981). Isolation and characterization of messenger RNAs for seed lectin and kunitz trypsin inhibitor in soybean. Plant Physiol 68:766.

Chemical Taxonomy, Molecular Biology, and Function
of Plant Lectins, pages 237–248

LECTINS AND LECTIN BINDERS (RECEPTORS) FROM PLANTS

Harold Rüdiger, Günter Gebauer and Johann Schurz
Institut für Pharmazie und Lebensmittelchemie der Universität, Am Hubland, D-8700 Würzburg

The biological role of lectins is still a matter of discussion. Experiments from different fields may contribute to solve this problem. In this context, the following questions will be discussed in this paper:
— What is the location of lectin?
— What can lectins and their fragments do *in vitro*, i. e. in an alien system?
— What do lectins effect *in vivo*, i. e. within their natural environment? The discussion about this question will be centered around the lectin binders (receptors), their interaction with lectins and with animal cells.

Leguminous lectins are mostly found in the cotyledones of the resting seeds. Only very little if any is found in other parts of the growing plant, depending on the species. In general, the small amounts outside the cotyledons are not at all equivalent to those which disappear from them during growth. It is unknown what happens to the lectins, whether they are degraded completely and unspecifically, or whether degradation products still may fulfil some functions.

Our own studies were made with the garden pea (*Pisum sativum*). As is seen from table 1, after germination the lectin content of the whole plant drops to a minimum. Outside the cotyledons, only very little lectin is found. During flower formation, however, a lectin increase is observed

Table 1. Lectin content at different growth states (μg/plant).

	0	9	18	39	69 days
total	680	433	18	337	10 237
cotyledons	--	420	5	--	--
above ground	--	6	4	333	10 226
subterraneous	--	7	9	4	11

first in the parts above ground, finally in the mature seeds. Again, outside the seeds the lectin content is very low, even in the pods which surround the seeds (not shown).
The localization of lectins within the cell is not yet clear. Cytoplasmatic as well as particle bound occurrence have been reported. We found the pea lectin in the soluble portion of the cell (scheme 1) using a centrifugation technique (Thomson et al. 1978; Van der Wilden et al. 1980).

Scheme 1. Intracellular localization of the pea lectin.

Extract
centrifugation
soluble fraction — 97% of the lectin
protein bodies — 3% of the lectin
density gradient centrif.
soluble fraction — 3% of the lectin
protein bodies — no lectin

In this, the seeds are imbibed in deionized water, then homogenized in the cold without adding ions. This keeps the protein bodies intact, i. e. they do not dissolve and can be centrifuged. In a following step, the protein bodies are sedimented in a discontinuous density gradient where they float on top of the bottom layer. After dialysis against buffer which leads to dissolution of the protein bodies, all fractions are analyzed for protein and lectin. As is seen from scheme 1, after the first step nearly all lectin is found in the supernatant, after the second purification step, also the rest of the lectin separates completely from the protein body fraction.

By immunochemical methods, several workers found the kidney and Jack bean lectins to reside in the cytoplasm (Mialonier et al. 1973; Clarke et al. 1975), on the other hand, association with protein bodies is reported (Bollini, Chrispeels 1978; Pusztai et al. 1979). This discrepancy may only reflect differences in the applied technique. Thus, possibly during preparation of microscopic objects, cytoplasmatic lectin may have been lost completely leaving behind only that bound to particles. Since fluorescence microscopy does not yield quantitative data, the soluble fraction might have been escaped the investigator's attention. It is well known that depending on the preparation technique, lectins may be found at sites where they are not in the living plant. On the other hand, possibly lectins are associated with the protein bodies but are so only very loosely. So they may already be released by water, possibly assisted by electrolytes from the plant material.

At any rate, lectins are inside the cell. Any suggested role outside will have to explain how the lectin migrates through the cell membrane and cell wall.

Besides agglutination, mitogenic stimulation of animal or human lymphocytes is another prominent capability of lectins. Most studies on mitogenicity have been performed with a particular lectin and various cell types, or it has been tried to find specific lectin receptors among the cell membrane proteins. Using a reverse approach, we tried to trace the mitogenic effect back to a particular structural feature of the lectin. In these experiments, we studied the lectin from Vicia sativa the fragments of which were at hand from the sequence determination (Gebauer et al. 1981). This lectin belongs to the two chain type (Foriers et al. 1979) and is very similar in its sequence to the pea, lentil and broad bean lectins. The Vicia sativa lectin is a typical T-cell mitogens as most lectins are. This is shown in fig. 1, where the action on mixed (B+T) cells is compared with that on B cells from nude mice. Moreover, its action is specifically suppressed by cyclosporin A, a fungal metabolite and T cell inhibitor (Bunjes et al. 1981).

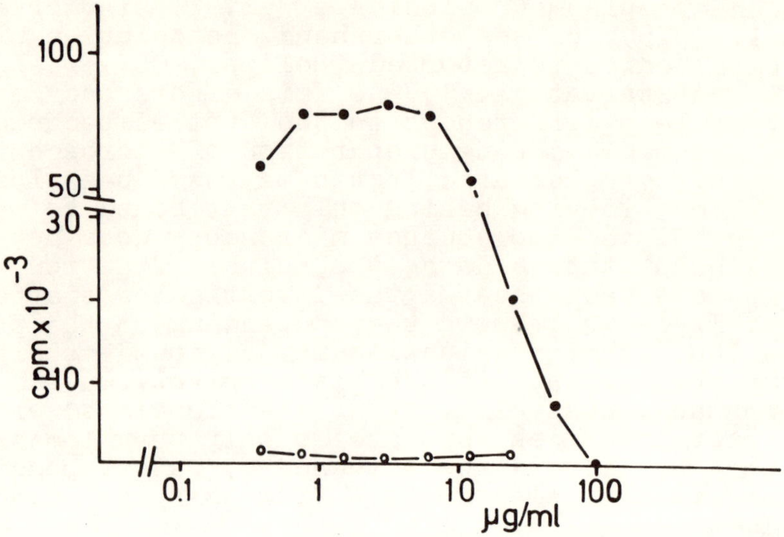

Fig. 1. Mitogenic stimulation, measured as ^{3}H-thymidine incorporation, by the Vicia sativa lectin. Splenocytes from Balb/c mice (B + T-cells): ●——●; splenocytes from nude mice (B-cells): o——o.

Isolated subunits or tryptic and chymotryptic peptides did not agglutinate red cells, but surprisingly their mitogenic activity was still retained. Table 2 summarizes these results. The ß-

Table 2. Relative mitogenic response of mouse splenocytes (Balb/c) to Vicia sativa lectin, its fragments and related peptides (medium = 1.0).

intact lectin	15.4	tryptic peptide T3 from α-chain	2.8
ß-chain	4.4	Ser-Trp	3.2
α-chain	3.0	Trp-Ser	3.5
α- and ß-chain	5.4		

chain acts as a mitogen at about 30% of the lectin's activity, the α-chain approximates 20%. If both chains are mixed, the effect is only additive which shows that they do not recombine. Even some of the peptides of the α-chain are mitogenic stimulators, the smallest one being active is the dipeptide seryl-tryptophan which is generated by chymotryptic cleavage from the positions 40/41 of the α-chain.

So it becomes clear that the mitogenic stimulation is a process which is related to the conformation not as closely as is the agglutination of red cells. In order to control this finding, we synthesized the dipeptide seryl-tryptophan and tested it. We obtained a 3.2fold stimulation at 10 - 50 µg/ml, which corresponds to a a quarter of the effect of the lectin. We compared this with the reverse sequence tryptophanyl-serine which does not occur as a proteolytic peptide but is also part of the sequence (positions 41/42). Also this peptide is a mitogen with a stimulation index of 3.5. The experiments with synthetical peptides invalidate the possible objection that the stimulation might have been achieved either by residual lectin or by bacterial lipopolysaccharides which often contaminate biochemical preparations.

In spite of all similarities between the lectin and its fragments in their mitogenic action, there are differences in the optimal concentrations. The lectin acts best at 5 µg/ml, subunits and peptides need much higher concentrations for optimal stimulation, mostly higher than was available to us. This means, that molar peptide concentrations necessary for stimulation are 300 - 400fold higher than of the lectin. This points to the important role of the protein conformation which brings the active sites to a position which is optimal for the cell surface receptors. The - chain contributes two active centers (seryl-tryptophanyl-seryl, pos. 40 - 42). The ß-chain sequence is known only partially but seems to be very similar to the corresponding sequences from the broad bean and lentil lectins. From these, two further active sites may be expected.

These considerations prompted us to immobilize the peptide because we hoped that this would possibly simulate the natural positioning better than in the freely dissolved state. Indeed, we obtained a significantly higher stimulation. As usual with large mitogens, the concentration-versus-stimulation curves are not bellshaped but run out in plateaus. This phenomenon is probably due to the fact that mitogen concentration high enough to suppress cell growth cannot build up. Table 3

Table 3. Relative mitogenic response of mouse lymphocytes (Balb/c) to an immobilized peptide.

Trp-Ser-Sepharose	14.0
Sepharose alone	2.2
medium alone	1.0

shows the result. Stimulation exceeds by far that by the dissolved peptide (table 2) or by the immobilizing carrier alone. Splenocytes from mice which are devoid of an lipopolysaccharide receptor are also significantly stimulated which shows that the effect is not due to contaminating lipopolysaccharides. Similarly enhanced mitogenic stimulation was obtained with the dipeptide which had been crosslinked with glutaraldehyde.

From these experiments, it appears that mitogenicity may be ascribed to relatively short sequence portions. Proper positioning and multiple presentation, however, appear to optimize the efficiency. The dipeptides which we found to be mitogenic are extremely amphiphilic since they are composed of a hydrophilic and a hydrophobic residue. Possibly their action mechanism resembles that of bestatin, a mitogenic and amphiphilic peptide from a Streptomyces, which is believed to act via a membrane protease (Müller 1981).

The sugar binding site of the lectin is not directly involved in mitogenicity. The glucose content of the lymphocyte culture medium is about 10 mM. Agglutination of erythrocytes, on the other hand, is already inhibited at 1 - 2 mM. For suppressing the mitogenic effect of the lectin by 50%, additional 35 mM have to be added which is not very far from inhibition by glucose in controls.

The effect of lectins within their natural environment, i. e. in plant cells, has been studies only scarcely. The problems in such experiments are that lectins have to be applied from outside whereas their natural location is inside. Moreover, if no effects are observed, this may only mean that a saturating amount of internal lectin is already present.

If, however, plants are supposed to be affected by lectins, they should also be expected to possess lectin receptors. We studied this question already some years ago (Gansera et al. 1979; Gebauer et al. 1979) and were surprised to find relatively homogeneous materials if we passed a plant seed extract over a column of lectin Sepharose. The subunit molecular weights of these lectin binding proteins (lectin receptors) are about 36 000 (*Canavalia ensiformis*, *Vicia faba*) and 52 000 (*Pisum sativum*, *Vicia sativa*). Molecular weights of the native proteins could only be determined with the last ones (*Pisum s.*, *Vicia s.*). These proteins are tetramers. The other lectin binders tend to form larger aggregates.

Lectins and lectin binders are different in size, in behavior towards red cells - the lectin binders are not agglutinins - and in amino acid composition. In contrast to the lectin, the lectin binder is distributed over the whole plant with a relatively large amount outside the cotyledons (table 4). Using the centrifugation technique mentioned above, we found also this protein in the water soluble fraction (scheme 2), though the adherence to protein bodies was somewhat stronger than with the lectin (scheme 1). Thus, apparently the pea lectin binder is a cytoplasmatic protein.

Table 4. Lectin binder content at different growth states (µg/plant).

	0	9	18	39	69 days
total	129	246	689	7 325	21 041
cotyledons	--	209	74	--	--
above ground	--	18	500	6 890	19 816
subterraneous	--	19	115	435	1 225

Scheme 2. Intracellular localization of the pea lectin binder.

Extract
centrifugation
soluble fraction — 75% of the lectin binder
protein bodies — 25% of the lectin binder
density gradient centrif.
28% of the lectin binder — soluble fraction
no lectin binder — protein bodies

The interaction between lectin and lectin binder is already manifested by the isolation procedure. We used some further methods in order to get insight into the binding parameters:
If a gel filtration column is equilibrated with the solution of a low molecular weight substance (in this case the lectin) and a solution of larger molecules (the lectin binder) passed over this column, an interaction between both species will be indicated by an increase in material at the elution volume of the high molecular weight substance and a corresponding trough at the position of the smaller one. The stoichiometry calculated from such experiments is about 4 pea lectin molecules per one lectin binder.
A further method which we used is laser nephelometry. This is a standard method for measuring immunoprecipitations. Fig. 2 shows the relative

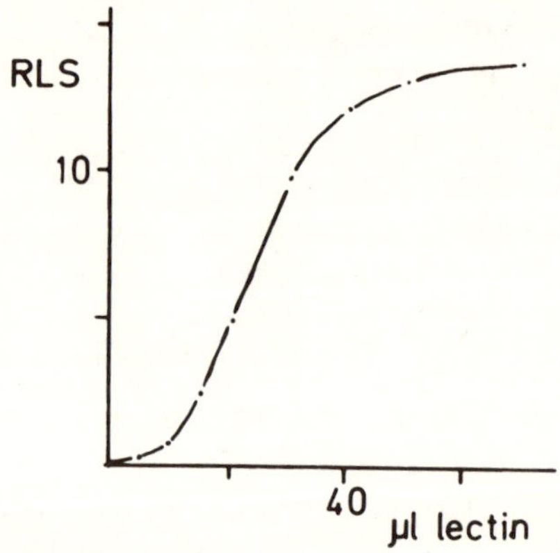

Fig. 2. Dependence of relative light scattering (RLS) on the amount of pea lectin added.
To 60 µg of lectin binder in 300 µl, rising amounts of lectin (2.2 mg/ml) were added in 0.05 M Tris/HCl pH 8, 0.1 M NaCl, 2% PEG 6000.

light scattering (RLS) if rising amounts of pea lectin are added to a fixed, predetermined concentration of the lectin binder. The sigmoidal curve may be interpreted in terms of cooperativity. From this, a Hill coefficient of 3.2 results. The dissociation constant of the lectin-lectin binder interaction, i. e. the half saturating concentration, is 3-5 µM. The association between

both proteins is very sensitive to glucose. 0.5 mM are sufficient to reduce the effect by 50%. High salt concentration and extreme pH values also weaken the interaction.
As a third method, we prepared a lectin binder column and studied the behavior of the lectin on it. Surprisingly, in normal buffer the lectin is only slightly retarded. This, however, can be increased, if the buffer is supplemented with sodium chloride. Thus, hydrophobic sites appear to participate in this interaction. Nevertheless, the presence of only little glucose totally abolishes any interaction, even at high sodium chloride concentrations.

Though lectin binders differ greatly from lectins, we tested if they at least might share with the lectins the ability to act as mitogenic stimulators. Table 5 compares their stimulation index with that of lectins. All eight proteins

Table 5. Relative mitogenic response of splenocytes from Balb/c mice (B + T-cells) to lectins and lectin binders (medium = 1.0)

	lectin	lectin binder
Canavalia ensiformis	17.7	6.2
Vicia faba	13.8	15.7
Vicia sativa	17.5	10.2
Pisum sativum	18.0	14.5

stimulate thymidine incorporation in typical bell-shaped dose-response-curves. With the lectins, the optimal stimulation occurs at 4 µg/ml, i. e. at relatively low concentrations, whereas an optimal stimulation by the lectin binders requires much higher concentrations, about 100 µg/ml. The effect which was achieved is, however, of similar size, in particular with the broad bean (Vicia faba) lectin binder which is the most active one. Stimulation by these proteins exceeds the effect of bacterial lipopolysaccharides by far.

Most lectins are T-mitogens. The standard mitogen for B-cells is a mixture of bacterial lipopolysaccharides which are chemically ill defined. So there exists a demand of B mitogens of known structure. The most simple method for studying B-cells is to make use of splenocytes from nude mice. This mouse strain lacks a functional thymus and therefore also functional T-cells. Using these cells, we looked if the lectin binders might possibly differ from lectins in their cell specificity. As controls, we ran parallel stimulation experiments with the lectins which should not stimulate. Table 6 shows the exciting result. As expected, the lectins are without any

Table 6. Relative mitogenic response of splenocytes from nude mice (B-cells) to lectins and lectin binders (medium = 1.0)

	lectin	lectin binder
Canavalia ensiformis	0.5	15.0
Vicia faba	0.8	27.7
Vicia sativa	0.4	1.9
Pisum sativum	0.4	2.2

influence on thymidine incorporation. This evidences that our cell preparation is virtually free from T-cells. On the other hand, however, the lectin binders from broad bean and Jack bean not only stimulate thymidine incorporation but do so to about the same extent as shown in Table 5 where mixed cells were studied. The other lectin binders do not act as strong B-mitogens.

As controls, we tested T-cells which had been purified from the natural mixture. They did not respond to lectin binders. On the other hand, the lectin binder induced stimulation of mixed cells is not suppressed by cyclosporin A, a T-cell suppressor (Bunjes et al. 1981). Simulation of the effect by contaminating lipopolysaccharides was excluded by the positive response of splenocytes from mice which lack an lipopolysaccharide receptor (Gebauer et al. 1982).

The existence of lectin binders and their location together with lectins inside the cell show that lectins are associated to other cell components. The specifity of both proteins towards animal cells may reflect some unknown kind of complementarity which is also active within the plant. So, by having shown that lectins are accompanied by natural binding partners, we have not just solved the question about the biological role of lectins but rather extended it, and, at the same time, also made a positive answer more probable.

Bollini, R, Chrispeels, MJ (1978). Characterization and subcellular localisation of vicilin and phythemagglutinin, the two major reserve proteins of Phaseolus vulgaris. Planta 142: 291.

Bunjes, D, Hardt, C, Röllinghoff, M, Wagner, H (1981). Cyclosporin A mediates immunosuppression of primary cytotoxic T-cell responses by impairing the release of interleukin 1 and interleukin 2. Eur J Immunol 11: 657.

Clarke, AE, Knox, RB, Jermyn, MA (1975). Localization of lectins in legume cotyledons. J Cell Sci 19: 157.

Foriers, A, De Neve, R, Strosberg, AD (1979). Lectin sequences as a tool for chemotaxonomical classification. Physiol Vég 17:597.

Gansera, R, Schurz, H, Rüdiger, H (1979). Lectin associated proteins from the seeds of Leguminosae. Hoppe-Seyler's Z Physiol Chem 360: 1579.

Gebauer, G, Schiltz, E, Schimpl, A, Rüdiger, H (1979). Purification and characterization of a mitogenic lectin and a lectin-binding protein from Vicia sativa. Hoppe-Seyler's Z Physiol Chem 360: 1727.

Gebauer, G, Schiltz, E, Rüdiger, H (1981). The amino-acid sequence of the α-subunit of the mitogenic lectin from Vicia sativa. Eur J Biochem 113: 319.

Gebauer, G, Schimpl, A, Rüdiger, H (1982). Lectin binding proteins as potent mitogens for B-lymphocytes from nu/nu mice. Eur J Immunol 12: 491.

Mialonier, G, Privat, JP, Monsigny, M, Kahlem, G, Durand, R (1973). Isolement, propriétés physico-chimiques et localisation in vivo d'une phytohémagglutinine (lectine) de Phaseolus vulgaris. Physiol Vég 11: 519.

Müller, WEG (1981). Effect of bestatin on DNA and RNA metabolism in T-cells. In Umezawa, H (ed). "Small molecular immunomodifiers of microbial origin". Pergamon Press, p. 39.

Pusztai, A, Croy, RRD, Stewart, JS, Watt, WB (1979). Protein body membranes of *Phaseolus vulgaris* cotyledons: isolation and preliminar characterization of constituent proteins. New Phytol 83: 371.

Thomson, JA, Schroeder, HE, Dudman, WF (1978). Cotylodenary storage proteins in *Pisum sativum*. Austr J Plant Physiol 5: 263.

Van der Wilden, W, Herman, EM, Chrispeels, MJ (1980). Protein bodies of mung bean cotyledons as autophagic organelles. Proc Natl Acad Sci USA 77: 428.

Posters

Chemical Taxonomy, Molecular Biology, and Function of Plant Lectins, pages 251–252

MICROHETEROGENEITY OF THE PHA ISOLECTINS

Carl A.K. Borrebaeck and Cristina Glad

Department of Pure and Applied Biochemistry
University of Lund
S-220 07 Lund, Sweden

Phytohemagglutinin (PHA) from Phaseolus vulgaris plant is a tetrameric lectin supposedly composed of two different subunits, E and L, with a molecular weight of 33 100 and 31 600, respectively. The lectin would then exist as five different isolectins which apparently originates from all possible tetrameric combinations of the two subunits.

A charge heterogeneity of the different isolectins was observed when they were run on an agarose gel electrophoresis at pH 8.6. This technique separated each isolectin into approximately three bands. The same splitting of the five isolectins was observed upon ion exchange chromatography (SP-Sephadex, pH 6.0) of an isolectin mixture and 14-15 different fractions could be isolated using a flat salt gradient in the elution buffer. The microheterogeneity of the isolectins was proven not to be due to 1) interactions between isolectin and agarose or Sephadex gels, 2) phenolic or proteolytic modifications, 3) carbamylation. The metal ion content (Ca^{2+}, Mg^{2+}) of fourteen different subfractions obtained from the ion exchange chromatography revealed no pattern that could explain the observed heterogeneity. The isolectins were, furthermore, enzymatically deglycosylated using endo-ß-N-acetylglucosaminidase H or D and chemically deglycosylated using trifluoromethanesulfonic acid. Deglycosylated isolectin gave the same electrophoretic patter using agarose electrophoresis as native PHA indicating that the charge microheterogeneity was most probably not due to diferences in glycosylation.

Experiments were also performed to determine if differences in the primary amino acid sequence of the PHA subfractions could explain the heterogeneity. Proteolytic digestion of the isolectin fractions was made using Staphylococcus aureus V-8 protease (45 minutes at 37 ^{o}C) followed by electrophoretic peptide mapping in polyacrylamide gels containing 8M urea or SDS-8M urea. Peptides separated according to molecular weight revealed two L-subunit specific and three E-subunit specific peptide bands. When the peptides were separated according to charge four L-subunit specific and two E-subunit specific bands were obtained. No differences could, however, be revealed between the different subfractions containing homotetrameric combinations of subunits. We finally tested the subunits using isoelectric focusing and PAGE in 8M urea and preliminary results indicate that the charge microheterogeneity might be due to the precence of several different types of E and L subunits.

Chemical Taxonomy, Molecular Biology, and Function of Plant Lectins, pages 253–254

THE RELATION BETWEEN METAL ION CONTENT AND MITOGENICITY OF PHA-L_4 AS ILLUSTRATED BY AGAROSE ELECTROPHORESIS.

Cristina Glad and Carl A.K. Borrebaeck

Department of Pure and Applied Biochemistry
University of Lund
Lund, Sweden

Phytohemagglutinin (PHA) from Phaseolus vulgaris plant is a tetrameric metalloprotein supposedly composed of two different subunits, E and L. The lectin would then exist as five different isolectins which apparently originates from all possible tetrameric combinations of the two subunits.

The isolectin PHA-L_4 is a commonly used mitogen in cell biology, immunology and genetics where it has been used for receptor, cell and chromosomal studies. PHA-L_4 is, furthermore, dependent on its content of divalent cations such as Ca^{2+}, Mn^{2+} and Mg^{2+}, for the ability to bind to carbohydrate structures.
It was found that on agarose gel electrophoresis at pH 8.6 PHA-L_4 had none or only slow migration towards the anode, while PHA-E_4 migrated toward the cathode. However, if the two isolectins were treated with EDTA their migration changed, that is PHA-L_4 exhibited a fast and PHA-E_4 a slow anodic migration.

Six different preparation of PHA were studied by agarose electrophoresis and it was found that the amount of demetallized isolectin varied depending on the method used for preparation. Thus, PHA-L_4 and E_4 prepared using affinity and ion exchange chromatography (pH 6.0) (Leavitt et al. 1977) contained no detectable amount of demetallized isolectin. 30-70% demetallized isolectin was found in PHA-L_4 prepared using 47% EtOH precipitation, DEAE-cellulose (pH 8.0) and Sp-Sephadex (pH 4.9) ion exchange chromatography followed by gel filtration (Räsänen et al. 1973). Up to 90% of demetallized PHA-L_4 was, furthermore, found when an extraction step using 0.1M HCl was included (Räsänen et al. 1973). The latter preparation could not be remetallized upon treatment with divalent cations. PHA prepared by retardation affinity chromatography on a N-acetyl-D-glucosamine gel (Chu, A., E-Y Laboratories, Inc., San Mateo, CA., personal communication) contained, however, less than 20% demetallized protein. This lectin preparation was the only one that fully retained the ability to be remetallized.

The different PHA preparations were then tested for their ability to induce lymphocyte transformation. A correlation between the metal ion content, as estimated by agarose gel electrophoresis, and the mitogenicity could be determined. PHA-L_4 prepared by HCl extraction showed a very low stimulating activity even at high concentrations (50 µg/ml), whereas the amplitude of the stimulation response for the other L_4 preparations was approximately the same although the lectin concentration at which the maximum stimulation occured increased with decreasing amounts of metal ions.

The results points to a fast and simple way to evaluate a preparation of PHA isolectins with regard to the metal ion content and subsequently their mitogenic activity. It also shows that different ways of isolating PHA can irreversibly change the molecular structure to the degree where the molecule has lost its ability to refold and bind divalent cations.

References
Leavitt, RD, Felsted, RL, Bachur, NR (1977). J Biol Chem 252:2961.
Räsänen, V, Weber, TH, Gräsbeck, R (1973). Eur J Biochem 38:193.

Chemical Taxonomy, Molecular Biology, and Function of Plant Lectins, pages 255–258

SAINFOIN LECTIN STRUCTURE

Kenneth D. Hapner, Rozenn N. Kouchalakos and Ralph A. Bradshaw

Chemistry Department, Montana State University, Bozeman, Montana 59717 (KDH and RNK); Department of Biological Chemistry, School of Medicine, Washington University, St. Louis, MO 63110 (RAB). Current address (RNK, RAB): Department of Biological Chemistry, California College of Medicine, University of California, Irvine, CA 92717.

Sainfoin (Onobrychis viciifolia, var Eski) lectin is a leguminous seed glycoprotein composed of two identical non-covalently associated 26000 dalton subunits (Hapner and Robbins, 1979). The lectin specifically agglutinates trypsinized cat erythrocytes and shows carbohydrate binding specificity toward D-mannose, D-glucose and their alpha-glycosidic derivatives. The purified protein contains 4.2% carbohydrate as a hexasaccharide chain attached to the polypeptide by N-glycosidic linkage to a single asparaginyl residue (Namen and Hapner, 1979).

We have determined the complete amino acid sequence of sainfoin lectin. The structure was generated from Edman degradation analyses of intact protein and derived tryptic and thermolytic peptide fragments. Seven resulting segments of continuous sequence were aligned on the basis of homology to give the final structure as shown below (detailed manuscript submitted elsewhere).

The sainfoin lectin molecule contains one polypeptide chain of 236 amino acid residues with a calculated molecular weight of 26509 daltons. Amino and carboxyl terminal residues are respectively alanine and threonine. The one sulfur containing residue is located at cysteine 33. Valine and isoleucine are present in equal amounts at position 49 and the glycosylation site is asparagine 118.

AENTVSFDFSKFLSGQENLILQGDTVTDDSNRCLVLTRENNGRPVQDS$\genfrac{}{}{0pt}{}{V}{I}$G 50

RVLYQTPIHLWDKQIDKEASFETSFTFFIYRENINRGGDGITFFLAPTDT 100

QPKSGGGYLGIFKDAESNETVVAVEFDTFSNRWDPANSHIGINVNSVKSK 150
(CHO attached at N118)

ITTPWGLKNDYFTVTITYDATRSLSVSSFYRNKPDDIFTVKASVHLRDAL 200

PQWVRIGLSAATGDLVEQHRLYSWSFKSVLPLDSST 236

Sainfoin lectin shows sequence homology with other sequenced leguminous lectins, i.e., favin (Hopp et al., 1982), concanavalin A (Wang et al., 1975; Cunningham et al., 1975), soybean (Hemperley et al., 1982) and lentil lectin (Foriers et al., 1981). When the sequences of the five lectins are aligned for maximum homology, the percent identities with sainfoin lectin are, respectively, 42, 39, 41, and 40%. Twenty-two percent of the amino acid residues are identical in all five proteins (indicated by the overline) and 62% are identical in at least three of the homologous proteins (indicated by the underline). The sainfoin lectin sequence is circularly permuted relative to that of concanavalin A, a feature shared with the other homologous lectins (Hemperley and Cunningham, 1983). This circular permutation results in correspondence of sainfoin lectin amino acid residues 1-118 and 119-236, respectively, to concanavalin A positions 121-237 and 1-120.

Further inspection of the primary structures of the homologous lectins relative to the three-dimensional structure of concanavalin A (Becker et al., 1975; Reeke et al., 1975) shows generally high conservation of sequence regions involved in ordered structural features of concanavalin A. Conservation of these sequences suggests that the homologous lectins will show an overall structure similar to that of concanavalin A. Greatest conservation is seen among those amino acids associated with the metal binding sites and the hydrophobic cavity. The calcium and manganese ions in concanavalin A are bound by the side chains of glutamic acid 8, aspartic acids 10 and 19, asparagine 14, histidine 24 and the carbonyl oxygen of tyrosine 12 (Becker et al., 1975). These residues are identical in all five lectins with the exception of position 12 which is conservatively replaced by phenylalanine in all cases. Among the 15 residues making up the hydrohobic cavity, 11 are conserved as identical or similar amino acids and three of the remaining four amino acids are conserved in lectins other than concanavalin A.

Less quantitative inspection of the homologous sequences shows that both 'front' and 'rear' beta structure is conserved with the 'front' being conserved to a somewhat greater extent than the 'rear' structure. Similarly, the contact sites between monomers are highly conserved. In contrast, the dimer-dimer salt bridges and residues making up the saccharide binding site are not conserved. The absence of sequence conservation in these areas may perhaps be expected in view of observed differences in molecular weight and saccharide binding affinity among the various lectins (Goldstein and Hayes, 1978).

The sainfoin lectin amino acid sequence, which is also homologous, in addition to the lectins discussed here, with other partially sequenced lectins (Foriers et al., 1979), supports the view that all leguminous lectins are members of a family of related plant proteins whose structure has been highly conserved for retention of important but as yet unconfirmed functional roles.

ACKNOWLEDGEMENT

Supported in part by grants from the United States Department of Agriculture (616-15-76) and the United States Public Health Service (AM 13362) and is published as paper number J1437 from the Montana Agricultural Experiment Station.

REFERENCES

Becker J W, Reeke G N, Wang J L, Cunningham B A, Edleman G M (1975). The covalent and three-dimensional structure of concanavalin A, III. Structure of the monomer and its

interactions with metals and saccharides. J Biol Chem 250:1513.
Cunningham B A, Wang J L, Waxdel M J, Edelman G M (1975). The covalent and three-dimensional structure of Concanavalin A, II. Amino acid sequence of cyanogen bromide fragment F_3. J Biol Chem 250:1503.
Foriers A, de Neve R, Strasberg A D (1979). Lectin sequences as a tool for chemotaxonomical classification. Physiol Veg 17:597.
Foriers A, Lebrun E, Van Rapenbusch R, de Neve R, Strasberg A D (1981). The structure of the lentil (Lens culinaris) lectin. Amino acid sequence determination and prediction of the secondary structure. J Biol Chem 256:5550.
Goldstein I J, Hayes C E (1978). Lectins: carbohydrate-binding proteins. Adv Carbohyd Biochem 35:128.
Hapner K D, Robbins J E (1979). Isolation and properties of a lectin from sainfoin (Onobrychis viciifolia, Scop). Biochim Biophys Acta 580:186.
Hemperly J J, Becker J W, Cunningham B A (1982). The structure and biological activities of plant lectins. In Bradshaw R A, Hill R, Tang J, Liang C, Tsao T, Tsou C (eds): "Proteins in biology and medicine," New York: Academic Press, p 395.
Hemperly J J, Cummingham B A (1983). Circular permutation of amino acid sequences among legume lectins. Trends Biochem Sci 8:100.
Hopp T P, Hemperly J J, Cunningham B A (1982). Amino acid sequence and variant forms of favin, a lectin from Vicia faba. J Biol Chem 257:4473.
Namen A E, Hapner K D (1979). The glycosyl moiety of lectin from sainfoin (Onobrychis viciifolia, Scop). Biochim Biophys Acta 580:198.
Reeke G N, Becker J W, Edleman G M (1975). The covalent and three-dimensional structure of concanavalin A, IV. Atomic coordinates, hydrogen bonding and quarternary structure. J Biol Chem 250:1525.
Wang J L, Cunningham B A, Waxdal M J, Edelman G M (1975). The covalent and three-dimensional structure of Concanavalin A, I. Amino acid sequence of cyanogen bromide fragments F_1 and F_2. J Biol Chem 250:1490.

Chemical Taxonomy, Molecular Biology, and Function
of Plant Lectins, page 259–260

THE PATHWAY OF CONCANAVALIN A DEPOSITION

Eliot Herman

Department of Biochemistry
University of California
Riverside, California 92521

The determination of the physiological role of seed lectins has remained elusive. Possible roles for seed lectins must be constrained by the intracellular or extracellular site(s) in which the protein is present. A few lectins have been localized at the intracellular level utilizing electron microscopy. These observations have indicated that the protein storage vacuole, the protein body, is the intracellular site of deposition. How proteins such as lectins are deposited within the protein body and by which organelles has remained controversial. The functional role of lectins may occur prior to final deposition and lectins may play a role in the deposition process.

In order to examine the deposition process during seed development, the lectin, concanavalin A (Con A), was localized in the cotyledon of developing jack beans (Canavalia ensiformis) by immunocytochemistry. Tissue samples utilized for immunocytochemical analysis were also subjected to SDS-PAGE. Electrophoresis was performed to examine the accumulation of Con A during seed development and its relationship to the deposition of other proteins, such as the storage protein, canavalin. A developmental sequence of 0.2, 0.7, 1.2, 2.0, 2.8 and 3.0 g fresh weight seeds were prepared and analyzed. Con A is first detected at 0.7 g fresh weight, and rapidly increases thereafter, until the seeds are fully mature at about 3.0 g. The accumulation of Con A occurs during the period of protein body filling from 1.2 to 2.8 g fresh weight. The accumulation of Con A occurs simultaneously with the accumulation of storage protein. Immunocytochemical observations were undertaken with 1.2 and 2.0 g fresh weight seeds in order to encompass the developmental timing of Con A deposition and protein body filling. Three g fresh weight seeds were also examined in order to ascertain the distribution of Con A at seed maturity. Staining the epoxy sections with antibody specific to Con A, followed by a second antibody specific to rabbit IgG coupled to colloidal gold, resulted in the specific labeling of intracellular components of jack bean cells.

Examination of mature green seeds revealed a dense specific labeling over the protein bodies. The distribution of gold particles was uniform within an individual protein body, among protein bodies within a single cell, and in protein bodies from adjacent storage parenchyma cells. The endoplasmic reticulum was extensively decorated following immunocytochemical labeling. All visible endoplasmic reticulum profiles appeared to be labeled. The labeling was enriched over the lumen with only sparse labeling associated with the endoplasmic reticulum surface and ribosomes.

Developmental differences in the degree of labeling of the endoplasmic reticulum were readily apparent. The ER was densely labeled at the developmental stages of the beginning (1.2 g fresh weight) and midpoint (2.0 g fresh weight) of the filling of the protein bodies. SDS PAGE gel observations indicate that Con A is also deposited at this developmental stage, therefore the ER is densely labeled at the developmental stage in which Con A is accumulated. Immunocytochemical observations of cells during the active protein accumulation stage (1.2-2.0 g fresh weight) revealed numerous Golgi bodies in close proximity to the incipient protein bodies. All Golgi bodies within the storage parenchyma cells were labeled. Densely labeled Golgi vesicles at the trans face of the Golgi apparatus were observed in the cytoplasmic space between the Golgi apparatus and protein body. These observations were interpreted to indicate that Con A is first sequestered in the lumen of the endoplasmic reticulum, then transferred to and through the Golgi body and finally deposited within the protein body by Golgi secretion vesicles. Supported by NSF grant PCM 8205148 to Leland M. Shannon.

Chemical Taxonomy, Molecular Biology, and Function
of Plant Lectins, pages 261–264

PREDICTION OF THE THREE-DIMENSIONAL STRUCTURES OF PLANT LECTINS

Kenneth W. Olsen

Department of Biochemistry
The University of Mississippi Medical Center
Jackson, MS 39216

All of the known amino acid sequences of leguminous lectins can be aligned starting at the amino terminus, except that of Concanavalin A (Foriers *et al.*, 1979). However, residue 123 of Con A aligns with the first residue of the other lectins. Comparison of the complete sequence of Con A (Cunningham *et al.*, 1975) with the sequence of favin (Cunningham, *et al.*, 1979), lentil lectin (Foriers *et al.*, 1978) and sanfoin lectin (Kouchalakos and Hapner, 1982) demonstrate an unusual circularly permuted homology. This homology can be explained if these lectins contain three domains (Olsen, 1983). In this communication, the domains of these lectins are described and the three-dimensional structure of Con A is modified to predict the conformation of the other lectins.

On the basis of the observed sequence homologies, the predicted domains of Con A should be residues 1-69 (domain 1), 70-122 (domain 2) and 123-237 (domain 3). The domain termini are in turns connecting strands of antiparallel sheet. These regions fold into compact units which have specific functions. The function of domain 1 is principally metal binding. All of the residues that interact directly with Mn^{2+} and Ca^{2+} are found in this region of the protein (Hardman *et al.*, 1982). In addition, the first domain is involved in one of the two major subunit contacts (Reeke *et al.*, 1975). Domain 2 is primarily responsible for subunit contacts. One-fourth of the residues in this region interact with other subunits. Domain 2 also has many contacts with the other two domains and helps hold them together. Domain 3 is responsible for the remaining subunit contacts.

The three-dimensional structure of the other leguminous lectins can be predicted by modifying the one for Con A. To make the predicted structure the local conformations at the amino and carboxyl termini were changed to bring them close enough to form the new peptide bond. This was easy to do since both residues are on the surface and only 17 Å apart. When the appropriate peptide bonds are broken to form the new terminii, the structure does not need to be greatly modified. This prediction will be tested since the X-ray crystal structures of peanut agglutinin (Olsen and Miller, 1982) and pea lectin (Meehan et al., 1982) are being determined.

Lectins contain two adjacent metal ion binding sites near the carbohydrate binding site. The structure of the manganese and calcium sites of Con A have been studied in detail by Hardman and his coworkers (1982). The following table shows shows that the amino acid side chains involved in metal binding are identical in Con A, favin, sainfoin lectin and lentil lectin. Thus, the predicted structures of sites of these other lectins are the same as the sites in Con A.

Residues in Metal Binding Sites of Lectins

Con A	Favin	Sainfoin	Lentil
Glu - 8	Glu - 123B	Glu - 125	Glu - 121B
Asp - 10	Asp - 125B	Asp - 127	Asp - 123B
Asn - 14	Asn - 129B	Asn - 131	Asn - 127B
Asp - 19	Asp - 133B	Asp - 134	Asp - 131B
His - 24	His - 140B	His - 139	His - 138B
Ser - 34	Ser - 150B	Ser - 149	Ser - 148B
Asp - 208	Asp - 82B	Asp - 89	Asp - 82B

Con A, favin, sainfoin lectin and lentil lectin have similar sugar specificities. Hardman and Ainsworth (1976) have determined the structure of the Con A-α-D-mannopyranoside complex at 6 Å resolution. The residues in the sugar-binding site of Con A are compared with predicted residues in the sites of the other lectins in the table below. These positions are not nearly as conserved as those in the metal binding sites.

Residues in Sugar-Binding Sites of Lectins

Con A	Favin	Sainfoin	Lentil
Tyr - 12	Phe - 127B	Phe - 129	Phe - 125B
Asn - 14	Asn - 129B	Asn - 131	Asn - 127B
Asp - 16	Ala - 131B	Arg - 132	Ala - 129B
Leu - 99	Ala - 30A	Asp - 214	Ala - 30A
Tyr - 100	Glu - 31A	Leu - 215	Glu - 31A
Asp - 208	Asp - 82B	Asp - 89	Asp - 82B
Arg - 228	Ser - 102B	Gly - 107	Gap

At several points the sequence homology requires that a residue in Con A be skipped in order to maintain the alignment. This results in a deletion in the three-dimensional structure of the second lectin when compared to Con A. All of these deletions are on the outside of the molecule and can be easily accommodated into the three-dimensional structure.

In conclusion, the three-dimensional structure of Con A contains three domains and can be used to predict the structures of other leguminous lectins. The metal binding sites of favin, sainfoin lectin and lentil lectin are identical to that of Con A. Although they have similar monosaccharide specificities, the sugar binding sites of these four lectins are quite different. Finally, the deletions needed to align the four sequences can be easily accommodated into the three-dimensional structures.

Cunningham BA, Wang JL, Waxdal MJ, Edelman GM (1975). Covalent and three-dimensional structure of Concanavalin A II. Amino acid sequence of cyanogen bromide fragment F_3. J Biol Chem 250 : 1503.

Cunningham BA, Hemperly JJ, Hopp TP, Edelman GM (1979). Favin versus concanavalin A : circularly permuted amino acid sequences. Proc Natl Acad Sci 76 : 3218.

Foriers A, de Neve R, Strosberg AD (1979). Lectin sequences as a tool for chemotaxonomical classification. Physiol Veg 17 : 597.

Foriers A, de Neve R, Kanarek L, Strosberg AD (1978). Common ancestor for concanavalin A and lentil lectin: Proc Natl Acad Sci 75 : 1136.

Hardman KD, Ainsworth CF (1976). Structure of the concanavalin A-methyl α-D-mannopyranoside complex at 6 Å resolution. Biochem 15 : 1120.

Hardman KD, Agarwal RC, Freiser MJ (1982). Manganese and calcium binding sites of concanavalin A. J Mol Biol 157 : 69.

Kouchalakos R, Hapner KD (1982). Amino acid sequence of sainfoin lectin. Fed Proc 41 : 1180.

Meehan Jr EJ, McDuffie J, Einspahr H, Bugg CE, Suddath FL (1982). Crystal structure of pea lectin at 6 Å resolution. J Biol Chem 257 : 13278.

Olsen KW (1983). Prediction of three-dimensional structure of plant lectins from the domains of concanavalin A. Biochim Biophys Acta 743 : 212.

Olsen KW, Miller RL (1982). Crystallization and preliminary x-ray data for peanut agglutinin. FEBS Lett 145 : 303.

Reeke Jr GN, Becker JW, Edelman GM (1975). Covalent and three-dimensional structure of concanavalin A IV atomic coordinates hydrogen bonding and quaternary structure. J Biol Chem 250 : 1525.

Chemical Taxonomy, Molecular Biology, and Function of Plant Lectins, pages 265–267

BIOCHEMICAL AND BIOLOGICAL COMPARISON OF *LATHYRUS* LECTINS

Pierre Rougé

Laboratoire de Biologie cellulaire
Faculté des Sciences pharmaceutiques
31062 Toulouse (France)

The lectins from the Leguminosae are tetramers which can be divided into two groups :

- the one-chain lectins which are composed of a single polypeptide chain with a MW between 25 000-30 000, are widely distributed among the tribes of Leguminosae, with the exception of the *Vicieae* tribe.
- the two-chain lectins which are made of two different polypeptide chains : the light or α chain and the heavy or β chain whose MW are respectively 4 000-6 000 and 18 000-20 000. These lectins occur only in the tribe of *Vicieae*.

The comparison of extensive amino acid sequences of various lectins and their subunits indicates that the Leguminosae lectins are conservative in their amino acid sequences and that striking similarities exist between the two above mentioned groups of lectins. In addition, these data have led to attractive speculations concerning the changes of the genes involved in lectin synthesis during evolution (see A.D. Strosberg).

Because the Leguminosae lectins appear as highly conserved and evolutionarily related proteins, one can hypothesize that comparison of lectins from closely related species, e.g. belonging to the same genus, would enable us to study the evolution of proteins and their genes during the speciation process.

Eight lectins from the genus *Lathyrus* (*L. aphaca* L.,

L. articulatus L., *L. cicera* L., *L. hirsutus* L., *L. ochrus* (L.) DC., *L. odoratus* L., *L. tingitanus* L. and *L. vernus* (L.) Bernh.) belonging to the tribe of *Vicieae* have been chosen for this study. Here are reported preliminary results dealing with the comparison of their main biochemical and biological properties in order to check the extent of their phylogenetic relationships.

All *Lathyrus* lectins studied appear as two-chain lectins made of α and β chains. The MW of α chains ranges from 4 000 to 5 000 while the MW of β chains is comprised between 19 000 and 20 000. Assuming, by comparison with pea and lentil lectins, that *Lathyrus* lectins are tetramers gives calculated MW ranging from 46 000 to 50 000, which agree with the values obtained by gel chromatography. In addition, the SDS-PAGE analysis of lectins reveals the constant presence of a polypeptide chain having a MW comprised between 17 000 and 18 000 which could correspond to a lectin precursor.

The amino acid composition of both lectins and their subunits shows very striking similarities. Like most Leguminosae lectins and many plant proteins, *Lathyrus* lectins are relatively rich in Asp, Thr and Ser but seem to be devoid of sulfur-containing amino acids. Similarly, β chains are rich in Asp, Thr and Ser while α chains contain more Glu, Ala, Leu, Val and His but are low in Asp, Ile, Tyr and Lys by comparison with β chains. However, despite these similarities between lectins and their subunits, some few unexpected values are encountered, e.g. the very low Ile content of *L. aphaca* α chain, the low Ser content of *L. ochrus* β chain or the high Ile content of *L. odoratus* α chain. The dinitrophenyl-derivatives of amino acids of the two subunits indicate that β chains always contain Thr as N-terminal amino acid while the N-terminal amino acid of α chains is Val (*L. cicera, hirsutus, odoratus, tingitanus* and *vernus* lectins) or Ala (*L. aphaca, articulatus, ochrus* lectins). Peptide mapping performed after trypsin hydrolysis of lectins shows both common and different peptides, as indicated by their electrophoretic mobilities.

All lectins contain approximately 2 Mn2+ and 3 Ca2+ atoms per mole. In addition, *L. articulatus* and *L. tingitanus* lectins contain trace amounts of Zn2+. All *Lathyrus* lectins studied contain only 0.2% or less carbohydrate and amino sugars are not detected.

The isolated lectins are nonspecific and agglutinate equally well human red blood cells of different ABO groups, like most Leguminosae lectins. Their haemagglutinating activity is inhibited best by D-mannose, D-glucose and their α-glycosides, α-methyl-D-mannoside being the most potent inhibitor.

All the tested lectins crossreact with heterologous antibodies giving essentially reactions of identity and, similarly, antibodies against α and β subunits from one *Lathyrus* lectin react respectively with α and β subunits from the other *Lathyrus* lectins. These results clearly indicate that lectins and their subunits bear some common determinants.

In addition, similarities between the binding sites of the lectins are evidenced by their ability to interact with various human serum glycoproteins. Many glycoproteins such as C1-inactivator, haptoglobin, α2-macroglobulin, ceruloplasmin, haemopexin, transferrin and the main three classes of immunoglobulins (IgA, IgG and IgM) always interact with lectins. However, a few lectins interact with other serum glycoproteins and the discrepancies observed among the lectins would presumably indicate a slightly different specificity of their binding sites.

As expected, the comparison of the main biochemical and biological properties of the eight *Lathyrus* lectins studied reveals striking similarities. These data, which of course reflect homologous amino acid sequences, suggest that *Lathyrus* lectins have to be regarded as very conserved proteins. Therefore, it appears that the comparison of the amino acid and nucleic acid sequences of these lectins together with the study of their immunological relations would allow the stepwise study of protein and gene evolution during the speciation process. Another interest of this work, which is now in progress, lies in the fact that it would enable to construct a detailed phylogenetic tree at the genus level.

Chemical Taxonomy, Molecular Biology, and Function of Plant Lectins, pages 269–270

PHYTOHEMAGGLUTININ: BIOSYNTHESIS, TRANSPORT AND MODIFICATION OF THE CARBOHYDRATE SIDECHAINS

Maarten J. Chrispeels and Alessandro Vitale

Department of Biology, University of California, San Diego

La Jolla, CA 92093

Phytohemagglutinin (PHA) is the major seed lectin of the common bean Phaseolus vulgaris. In most varieties PHA accounts for 5-10% of the seed protein. Phytohemagglutinin is present primarily in the storage parenchyma cells of the cotyledons where it occurs in the protein bodies together with phaseolin, the major seed storage protein.

The synthesis of PHA starts with the biosynthesis of the polypeptide chains by polysomes attached to the endoplasmic reticulum (ER). Insertion into the lumen of the ER is accompanied by two co-translational events: the removal of a signal peptide and glycosylation with high-mannose oligosaccarides. Glycosylation is inhibited by tunicamycin and presumed to be mediated by dolicholpyrophosphate.

When developing cotyledons are incubated with radioactive fucose, PHA is the most abundant labeled protein. The incorporation of fucose into the carbohydrate moiety of PHA occurs in the Golgi apparatus. Analysis of the oligosaccharide sidechains of PHA which has accumulated in the protein bodies indicates that there are at least two kinds of sidechains. These chains can be separated as glycopeptides on Biogel P4 columns after exhaustive digestion of the PHA with pronase and proteinase K. One kind of chain is of the high-mannose type and the other kind is of the complex type with fucose and xylose. The former is susceptible to degradation by α-mannosidase and endo-β-N-acetylglucosaminidase, while the latter is resistant to both enzymes. Each polypeptide of PHA has one high-mannose and one complex oligosaccharide sidechain, indicating that only one of the two high-mannose sidechains initially formed in the ER is modified in the Golgi complex.

Transport of PHA from the Golgi complex to the protein bodies is mediated by small (0.1-0.2μ) electron-dense vesicles filled with storage protein. Arrival of PHA in the protein bodies is followed by a slow increase in the electrophoretic mobility of the glycopolypeptides of PHA on SDS-polyacrylamide gels. This decrease in the apparent molecular size is not due to a change in the polypeptide moiety, but to a slow decrease in the size of the complex oligosaccharide chains. The decrease involves the removal of peripheral N-acetylglucosamine residues which were attached to the complex sidechains in the Golgi complex. This slow removal is thought to be carried out by β-N-acetylglucosaminidase located in the protein bodies. The sodium ionophore monensin prevents the transport of PHA to the protein bodies and the processing of the complex oligosaccharide chains.

Chemical Taxonomy, Molecular Biology, and Function of Plant Lectins, pages 271–272

LOCAL AND SYSTEMIC RESPONSES TO DIETARY LECTINS

A. Pusztai*, F. Greer*, M. de Guia Silva Lima†, A. Prouvost-Danon≠ and T. P. King*

*The Rowett Research Institute, Bucksburn, Aberdeen, Scotland, U.K.

†Biochemistry Department, Universidade Federal Do Ceara, 60,000 Fortaleza (CE), Brazil

≠Immunotherapie Experimentale, Institut Pasteur, 75015 Paris, France

Inclusion of pure *Phaseolus vulgaris* lectins (or beans) in diets for rats, mice, pigs or steers increased both faecal and urinary losses of N and resulted in a negative N balance for the animals.[1-3] Apart from relatively minor effects on organs such as the liver, pancreas and spleen, the main target tissues of the lectin were the thymus and the small intestine. The weight, the cellular content (protein and carbohydrate) and cell number (DNA) of the small intestine were increased by as much as 60-80% in about 10 days. The effect was reversible within certain limits. At the ultrastructural level severe disorganization of the brush border of the proximal small intestine was observed with disruption of the microvilli of the absorptive enterocytes.[4] Immunocytochemical studies have provided evidence that this lesion is associated with the binding of lectins to the luminal surfaces of enterocytes.[5] It is suggested that this results in an interference with the intermediate and final stages of protein breakdown and N absorption in the gut. However as the animals were in a negative N balance the results also indicated the existence of catabolic N losses due to systemic effects. Animals on lectin-containing diets developed circulating antibodies (IgG-class) of high titre to the lectin but not to other components of the diet. Moreover, there was a direct relationship between the severity of the toxic symptoms and the titre of the antibody. These and direct studies on the uptake of ^{125}I-lectin from the gut suggested that a relatively large proportion (1-5%) of the lectin traverses the intestinal barrier and gains access to systemic circulation. Thus by inference, these animals had an inadequate concentration of lectin-specific s-IgA in the gut. In addition mice (and rats) were shown to develop an immediate-type hypersensitivity reaction due to lectin-specific IgE. These results suggest that lectin toxicity is due to the combined effects of the interference with intestinal digestion/absorption of N through the damaged enterocytes and to local and systemic immune (and other) responses of the animals to the internalised lectin.

REFERENCES

1. A. Pusztai, E.M.W. Clarke and T.P. King (1979) Proc. Nutr. Soc. 38, 115-120.
2. A. Pusztai, E.M.W. Clarke, G. Grant and T.P. King (1981) J. Sci. Food Agric. 32, 1037-1046.
3. A. Pusztai, T.P. King and E.M.W. Clarke (1982) Toxicon 20, 195-197.
4. T.P. King, A. Pusztai and E.M.W. Clarke (1982) J. Comp. Path. 92, 357-373.
5. T.P. King, A. Pusztai and E.M.W. Clarke (1980) Histochem. J. 12, 201-208.

Chemical Taxonomy, Molecular Biology, and Function of Plant Lectins, pages 273–276

LATERAL DIFFUSION OF LECTIN RECEPTORS ON THE PLASMA MEMBRANE OF SOYBEAN CELLS

John L. Wang, Thomas N. Metcalf, III, and Melvin Schindler

Department of Biochemistry
Michigan State University
East Lansing, MI 48824

Protoplasts prepared from suspension cultures of root cells of Glycine max (SB-1 cell line) (Constabel 1975) bound soybean agglutinin (SBA), concanavalin A (Con A), and wheat germ agglutinin (WGA). Binding studies carried out with ^{125}I-labeled SBA, Con A, and WGA showed that these interactions were saturable and saccharide-specific. Fluorescence microscopy demonstrated uniform membrane labeling by the fluorescein-derivatized lectins. Calcofluor staining and scanning electron microscopy showed neither the characteristic fluorescence indicative of cell wall material nor cellulose microfibrils, respectively. These results suggest that the observed lectin binding was due to the interaction of the ligand with oligosaccharide moieties on the plasma membrane.

The lateral mobility of fluorescein-labeled lectins bound on protoplasts at 23° was determined using the technique of fluorescence redistribution after photobleaching (FRAP) (Koppel et al 1981). The diffusion constant for SBA, at a concentration of 50 μg/ml, was 4.6 x 10^{-11} cm^2/sec (Table 1). We did not obtain any significant change in the diffusion coefficient when the photobleaching experiments were repeated using either very low (5 μg/ml) or very high (250 μg/ml) concentrations of SBA. The presence of sodium azide in the protoplast medium was shown to have no effect on the mobility of SBA bound to protoplasts. In addition, a series of photobleaching experiments was conducted in the presence of coumarin, an inhibitor of cell wall synthesis (Meyer, Herth 1978). The resulting diffusion coefficient showed no significant difference from the data of Table 1. Thus, the slow lateral mobility of SBA bound to protoplasts is not due

Table 1 - Diffusion Coefficients of Lectins Bound to Soybean Protoplasts at 23°

Probe*	Treatment	D** ($cm^2/sec \times 10^{10}$)	% Recovery **
SBA	--	0.46 ± 0.21	91 ± 6
Con A	--	0.72 ± 0.52	78 ± 11
WGA	--	3.0 ± 1.8	84 ± 11
WGA	SBA (50 μg/ml)	0.46 ± 0.17	34 ± 11
WGA	colchicine (1 uM) SBA (50 ug/ml)	1.3 ± 0.91	37 ± 11
WGA	lumicolchicine (1 uM) SBA (50 ug/ml)	0.35 ± 0.06	36 ± 9

* all experiments were done using 50 ug/ml of fluorescently labeled lectin.

** values are expressed as mean ± standard deviation.

to regeneration of cell wall components and is not affected by lack of metabolic energy.

The diffusion constants of two other lectins bound to their receptors on soybean protoplasts were also determined. Con A yielded a D value of 7.2×10^{-11} cm^2/sec, similar to that of SBA (Table 1). In contrast, WGA gave D values of 3 x 10^{-10} cm^2/sec (Table 1). Therefore, it was apparent that lectins bound on soybean membranes exhibited two rather arbitrary classes of lateral mobility: (a) fast ($D = 3 \times 10^{-10}$ cm^2/sec) and (b) slow ($D = 5 \times 10^{-11}$ cm^2/sec).

Perhaps the most striking observation in the present study is the effect of SBA and Con A on the mobility of WGA receptors on the soybean plasma membrane. Pretreatment of the protoplasts with either SBA or Con A, followed by incubation with fluorescein-derivatized WGA, resulted in a six-fold reduction in the diffusion coefficient of the latter lectin ($D = 4.6 \times 10^{-11}$ cm^2/sec; Table 1). This effect cannot be ascribed to the anchoring of one set of relatively mobile receptors to another set of immobile receptors through lectin-lectin interactions. This conclusion is based on gel filtration experiments which suggest that WGA does not interact with SBA or Con A. Moreover, the binding of SBA does not change the amount of WGA bound to the soybean cell. Therefore, the decrease in D values of WGA in the presence of SBA most probably cannot be due to changes in the number or population of receptors. Finally, preliminary experiments indicate that SBA does not share the same receptors with Con A and WGA on soybean cells. Therefore, it appears that the binding of SBA and Con A results in alterations of other components of the soybean plasma membrane in such a way as to restrict the mobility of the WGA receptors. To the best of our knowledge, this is the first report of an endogenously-produced protein, SBA, which is capable of modulating the dynamic properties of its own membrane.

Although the detailed mechanism of modulation remains to be investigated, the present results may be analogous to the modulation by Con A of receptor mobility in a variety of animal cells (Edelman et al 1976). The binding of Con A to lymphocytes inhibits patch and cap formation of cell surface immunoglobulin as well as many other different receptors. It has also been shown that Con A binding results in a seven-fold reduction in the D values of surface immunoglobulin of lymphocytes (Henis, Elson 1981) and a ten-fold reduction in the D values of receptors on 3T3 mouse fibroblasts (Schlessinger et al 1977). This modulation can be partially reversed by colchicine, implicating a role for microtubules in the Con A effect. Consistent with this proposed analogy, we have found that colchicine also partially reversed the effect of SBA on the mobility of WGA receptors ($D = 1.3 \times 10^{-10}$ cm^2/sec; Table 1). Lumicolchicine, a photo-inactivated derivative of colchicine that does not bind to tubulin (Wilson, Friedkin 1967), had no effect on the SBA modulation. These results strongly suggest that the binding of external ligands to the plasma membrane of plant cells can alter the cytoskeletal structures of these cells. These alterations may be similar to animal cell membrane rearrangements which occur in the process of ligand-receptor mediated endocytosis. The functional significance of this type of membrane change for a plant cell may be in the binding and subsequent transport of Rhizobium across the membrane of root hair cells to start the process of nodule formation.

Acknowledgments: This work was supported by grants PCM-8011736 from NSF, GM-27203 from NIH and a McKnight Award For Individual Research Projects in Plant Biology.

References

Constabel F (1975). Isolation and culture of plant protoplasts. In Gramborg OL, Wetter LR (eds): "Plant Tissue Culture Methods", Saskatoon, Saskatchewan, Canada: National Research Council of Canada, p 11.

Edelman GM, Wang JL, Yahara I (1976). Surface modulating assemblies in mammalian cells. In Pollard T, Goldman R, Rosenbaum J (eds): "Cell Motility", Cold Spring Harbor, New York: Cold Spring Harbor Laboratory Press, p 305.

Henis YI, Elson EL (1981). Inhibition of the mobility of mouse lymphocyte surface immunoglobulins by locally bound concanavalin A. Proc Natl Acad Sci USA 78:1072.

Koppel DE, Sheetz MP, Schindler M (1981). Matrix control of protein diffusion in biological membranes. Proc Natl Acad Sci USA 78:3576.

Meyer Y, Herth W (1978). Chemical inhibition of cell wall formation and cytokinesis but not nuclear division in protoplasts of Nicotiana tobacum L. cultivated in vitro. Planta 142:253.

Schlessinger J, Elson EL, Webb WW, Yahara I, Rutishauser U, Edelman, GM (1977). Receptor diffusion on cell surfaces modulated by locally bound concanavalin A. Proc Natl Acad Sci USA 74:1110.

Wilson L, Friedkin M (1967). The biochemical events of mitosis. II. The in vivo and in vitro binding of colchicine in grasshopper embryos and its possible relation to inhibition of mitosis. Biochemistry 6:3126.

Chemical Taxonomy, Molecular Biology, and Function of Plant Lectins, pages 277–278

A LATEX AGGLUTINATION ASSAY FOR LECTIN-GLYCOPROTEIN INTERACTIONS

Sándor Pongor[a] and Zoltán Riedl[b]

[a]The Rockefeller University, New York 10021

[b]HUMAN, Institute for Serobacteriological Research Budapest, Hungary

The most extensively used methods for the study of lectin-mediated interactions are based on cell agglutination phenomena. Interpretation of cell agglutination patterns is hampered, however, by the facts that (a) nonspecific agglutination can occur as a result of ionic or hydrophobic interactions (Kauss, this volume); (b) cells may undergo lysis; and (c) lectins may simultaneously interact with different moieties present on cell surfaces. To overcome these difficulties, we designed a latex agglutination test, in which polystyrene latex particles, coated with a glycoprotein - and second-coated with an inert protein, usually BSA - are used to detect lectin activity (Pongor and Riedl 1983). Blood group A+H substance (BGA) was fixed to polystyrene latex (0.3-3 microns) by physical absorption. Agglutination of this latex by Con A, SBA and WGA is specifically inhibited by the respective monosacharide haptens (20 mM) and is not affected by the known noninhibitory sugars. The sensitivity of the assay depends on particle loading and latex concentration, under optimal conditions it approaches that of the red cell agglutination assay. NaCl, $NaHCO_3$, NH_4-acetate and Tris.HCl up to a concentration of 1M do not interfere with the test. This makes it possible to minimize nonspecific agglutination, without the risks of cell lysis. BSA-coated latex may serve as an appropriate controll for this purpose.

Another application of the principle is to use lectin-coated latex particles to detect glycoproteins in solution. Con A - latex e.g. is agglutinated by maltosylated BSA (prepared by reductive alkylation) provided that the molecule contained more than on carbohydrate residue that were simultaneously accessible for the lectin. This makes it possible in certain cases to distinguish between molecule populations that differ in the degree (and/or topology) of glycosilation.

Reference:

Pongor S and Riedl Z (1983). A latex Agglutination Test for Lectin Binding. Anal. Biochem. 129: 51

Chemical Taxonomy, Molecular Biology, and Function
of Plant Lectins, pages 279–280

STRUCTURE AND EXPRESSION OF PHASEOLUS VULGARIS LECTIN GENES

Leslie M. Hoffman

Agrigenetics Advanced Research Laboratory,
5649 East Buckeye Road
Madison, Wisconsin 53716

Lectins belong to one of two groups of major seed glycoproteins in Phaseolus vulgaris, the French bean. Using poly(A) RNA from bean cotyledons, a cDNA library was constructed in plasmid pBR322. The library was screened with a cloned pea lectin cDNA (kindly provided by Dr. T. J. V. Higgins) to yield six hybridizing clones. The cDNA of one plasmid, pPVL134, was subjected to nucleotide sequence analysis and encodes a 223 amino acid lectin and a presumptive signal peptide of 20 or 22 residues.

The 5' region of pPVL134 contains four ATG codons, the first of which is followed by a termination codon after three residues. The second ATG initiates a 58 amino acid peptide containing almost 25% histidine, lysine plus arginine residues. The third and fourth ATG codons are separated by one codon and initiate the lectin plus signal peptide.

The expression of lectin genes in various tissues at several stages of development was examined by Northern blot hybridization of total RNA with cloned lectin cDNA probe. One kb long lectin transcripts begin accumulating in cotyledons by the 9mm stage, increase in quantity until about the 16mm stage, then decrease dramatically until seed desiccation. Larger hybridizing RNA species are consistently detected at a lower abundance than the 1 kb transcripts.

Lectin-specific transcripts are also found in leaves of plants bearing pods, but not in young leaflets or roots. Both leaf and seed lectin RNAs are polyadenylated. In a membrane fraction of mature bean leaves, a polypeptide has been identified by Western blotting which cross reacts with antiserum to seed lectins. The leaf protein migrates faster on an SDS gel than the seed lectins.

From a bacteriophage λ1059 library of P. vulgaris DNA several lectin genomic clones were selected using pPVL134 cDNA as a probe. One clone, λA9, has been shown by DNA sequence analysis to encode the mRNA from which pPVL134 was derived. The gene contains no intervening sequences and is flanked by regions exceptionally high in A+T content (72-75%). Potential transcriptional regulatory sequences such as the "TATAAA" and "CAAAT" boxes are found in their normal positions 5' to the transcriptional start site. The 5' untranslated region of the mRNA is only 10-16 nucleotides long, as determined by S_1 nuclease mapping.

I wish to determine whether the protein encoded by the cloned gene is a hemagglutinating or mitogenic lectin and whether it requires glycosylation for biological activity. To this end, I have expressed the lectin cDNA in E. coli maxicells under lac operator-promoter control and have constructed "tac" promoter expression plasmids containing the chromosomal gene. Tests for the agglutinating and mitogenic properties of bacterially synthesized lectin will be performed.

Chemical Taxonomy, Molecular Biology, and Function of Plant Lectins, pages 281–288

INTERACTION OF MOUSE CYTOTOXIC T CELLS WITH VARIOUS LECTINS

Toshiaki Osawa, Takeshi Yamazaki and Yasuyuki Imai
Division of Chemical Toxicology and Immunochemistry, Faculty of Pharmaceutical Sciences, University of Tokyo, Tokyo 113, Japan

The application of lectins for the separation of biologically distinct cell subpopulations has been widely investigated by many workers. Peanut agglutinin (PNA), for example, has been used for the characterization of immature thymocytes (Reisner et al. 1976), and for the concentration of concanavalin A (Con A)-induced mouse suppressor T cells (Imai et al. 1979; Nakano et al. 1980). Surface markers specific for cytotoxic T lymphocytes (CTLs) have also been investigated using lectins by Kimura et al. (1978; 1979). They found that Vicia villosa lectin (VVL) exhibited selective reactivity for the T145 glycoprotein that is expressed specifically on the cell surface of CTLs (Kimura, Wigzell 1978). However, other workers have recently reported that the binding site for VVL cannot be regarded as an exclusive marker for CTLs (Kaufmann, Berke 1981; MacDonald et al. 1981; Lang et al. 1982).

In this study, mouse cytotoxic T lymphocytes (CTLs) induced in vivo and in vitro in mixed lymphocyte cultures (MLC) were fractionated into agglutinated cells and unagglutinated cells by use of various lectins according to the method described previously (Nakano et al. 1980). Cytolytic activity was enriched in the agglutinated cell fraction with various 2-acetamido-2-deoxy-α-D-galactopyranose (GalNAc)-specific lectins, namely, Dolichos biflorus agglutinin (DBA), Helix pomatia agglutinin (HPA) and Phaseolus limensis agglutinin (LBA) as shown in Fig. 1. Only a little cytolytic activity remained in the unagglutinated cell fraction. Furthermore, the relationship between the binding of lectins and the cytolytic activity in various CTL cell lines was investigated using a fluorescence activated cell sorter (FACS II).

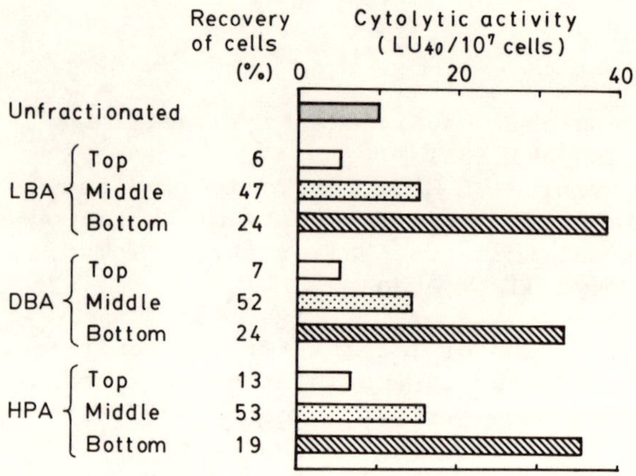

Fig. 1. Cytolytic activity of lectin-fractionated lymphocytes primed with P815 in vivo. One unit (LU_{15}) is defined as the number of lymphocytes required to lyse 15% of 2 x 10^4 P815 cells as target cells in a 4-h ^{51}Cr-release assay.

These interleukin-2-dependent CTL lines were long-term culture blast cell lines orginating from MLC between C57BL/6 (responder) and DBA/2 (stimulator), and their cytolytic activities are listed in Table 1. The binding of DBA to the cell lines was analyzed by a FACS II after staining with biotinyl DBA and fluorescence isothiocyanate (FITC)-labeled avidin. The results are shown in Fig. 2. Normal spleen cells and normal T cells showed no significant fluorescence. However, secondary MLC cells and CTL cell lines showed brighter fluorescence. The proportion of bright cells and the fluorescence intensities at the peaks in the fluorescence profiles increased almost parallel to the cytolytic activity. The same analysis was carried out using various lectins, and the relevance of the lectin binding to the cytolytic activity of CTL cell lines was investigated by calculating the Spearman rank correlation coefficients. As shown in Fig. 3, high rank correlation was

Table 1. Cytolytic activities of IL-2-dependent CTL lines*

Cells	Cytolytic activity ($LU_{50}/10^6$ cells)
Con A-activated spleen lymphocytes	0.036
Spleen lymphocytes primed *in vivo*	0.21
Primary MLC cells	1.8
Secondary MLC cells	4.8
LC5(-f) cells	6.0
LC7 cells	6.9
LC4 cells	25
LC5	192
$LC5T^8$	152
$LC5B^5$	227

* LC4, LC5, LC7 and LC(-f). LC4 and LC7 were long-term culture blast cell lines originating from secondary MLC between C57BL/6 (responder) and DBA/2 (stimulator), and LC5 originated from tertiary MLC in the same system. LC4 and LC5 cells were successively cultured in IL-2-supplemented medium in the continuous presence of mitomycin-C-treated stimulator cells. LC7 cells were cultured in the absence of stimulator cells. In the case of a portion of the LC5 cells, stimulator cells were removed 7 weeks after the start of the culture and thereafter they were maintained only in the presence of IL-2. This cell line was designated as LC5(-f). The cytolytic activity of these cell lines against P815 was tested. One unit (LU_{50}) is defined as the number of lymphocytes required to lyse 50% of 2 x 10^4 target cells in a 4-h ^{51}Cr-release assay.

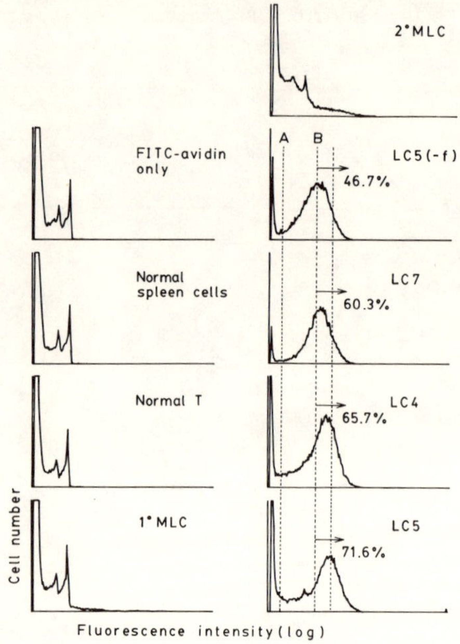

Fig. 2. Fluorescence profiles of normal cells and CTL cell lines labeled with biotinyl DBA and FITC-avidin. The values shown as percentages represent the ratios:

$$\frac{\text{(the area having a higher fluorescence intensity than B in a peak)}}{\text{(the area having a higher fluorescence intensity than A in a peak)}} \times 100$$

found between the cytolytic activity of these CTL cell lines and the binding of DBA or HPA to them, but weaker rank correlation in the cases of LBA and peanut agglutinin (PNA). Actually, DBA was found to bind to almost all cells of interleukin-2 (IL-2)-dependent CTL cell lines tested.

These results suggest that CTLs have abundant terminal α-D-GalNAc groups in the sugar chains on the cell surface. However, this correlation was found to be not absolute, because a minor proportion of the strong CTL cells was not stained with DBA.

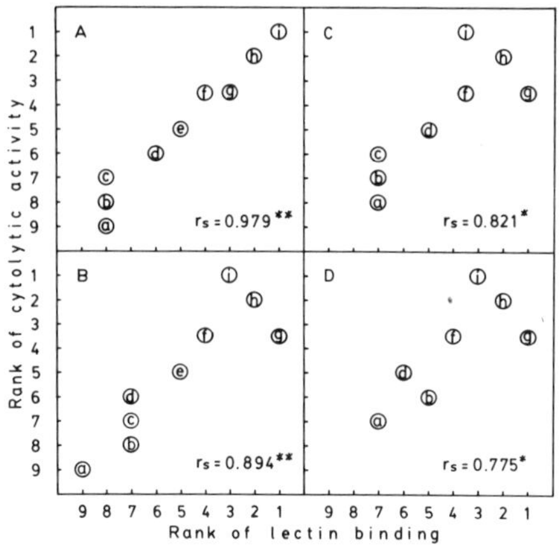

Fig. 3. Correlation between cytolytic activity and lectin binding in CTL populations. The binding of four kinds of lectins, DBA (A), HPA (B), LBA (C) and PNA (D), to various CTL cell lines was analyzed with a FACS II. The ranks of lectin binding were determined on the basis of fluorescence profiles. ** and * mean that the of r_s (Spearman rank correlation coefficient) values are significant at levels of 1% and 5%, respectively. CTL populations: a, normal T cells; b, Con A-activated spleen lymphocytes; c, spleen lymphocytes primed with P815 in vivo; d, primary MLC cells; e, secondary MLC cells; f, LC5(-f) cells; g, LC7 cells; h, LC4 cells; i, LC5 cells.

Thus, LC5 cells, which had the strongest cytolytic activity, were fractionated by the agglutination-sedimentation method with DBA. The major part (90%) of the cells was recovered in the agglutinated fraction and a much smaller part (3%) of the cells was found to be not agglutinated. However, the difference in cytolytic activity was not so marked between these two cell populations. Then, both the cell populations obtained by DBA-fractionation of LC5 cells were recultured in fresh medium, and the selection, namely DBA-fractionation and reculturing, was repeated several times for both cell populations to further purify the DBA-agglutinated cells and DBA-unagglutinated cells. The cytolytic activity of unagglutinated cells ($LC5T^8$ cells:

The selection was repeated 8 times) and agglutinated cells (LC5B^{5} cells: The selection was repeated 5 times) thus obtained was not so different from that of the original LC5 cells (see Table 1). These results shows that a minor groups of CTLs has no binding sites for DBA. Therefore, our data suggest that the binding sites for α-D-GalNAc-specific lectins cannot be regarded as an exclusive differential marker for CTLs. However, since the major part of CTLs can be agglutinated by DBA, its use may have practical value for the enrichment of CTLs.

It is of interest to note that the highest rank-correlation between the cytolytic activity of CTLs and the lectin binding to them was observed in the case of DBA among the α-D-GalNAc-specific lectins employed in this study. DBA is known to be the most specific lectin toward α-GalNAc residues and its most potent inhibitor is a blood group A active pentasaccharide, GalNAcα1→3(Fucα1→2)Galβ1→4GalNAc(1→6)R (Etzler, Kabat 1970). The other lectins used, HPA and LBA, appear to be less specific toward α-GalNAc residues (Galbraith, Goldstein 1972; Hammarström 1972) and may also react, to a more limited extent, with terminal α-Gal residues. It may, therefore, be reasonable to assume that this wider sugar-binding specificity of the latter two lectins is reflected in their lower selectivity for CTLs. Soy bean lectin (SBA) is also a specific lectin toward α-GalNAc residues (Lis *et al*. 1970; Pereira *et al*. 1974), but an L-fucosyl residue in the blood group A active pentasaccharide is known to greatly diminish the inhibitory activity of the oligosaccharide toward SBA. Since SBA is ineffective in the separation of CTLs (data not shown), most CTLs may express a blood group A antigen-like structure including an L-fucosyl residue and DBA can bind most specifically to this sugar structure. In this context, it is also of interest to note that the blood group A active pentasaccharide is much less active as an inhibitor toward VVL than GalNAcα1→3Gal (Kaladas *et al*. 1981). Although VVL was not tested in this study owing to its unavailability, it may be less selective for CTLs than DBA. Membrane receptors of CTLs for DBA may not exactly be a blood group A antigen itself, because immune anti-human blood group A serum used in this study could not react with CTLs. Another possibility could be that the anti-A antiserum recognize the larger portion of the blood group A antigen. The characterization of membrane receptors of CTLs for DBA, even if they are not exclusive differential cell surface markers for CTLs, is of interest and constitutes the next subject of our studies.

Etzler ME, Kabat EA (1970). Purification and characterization of a lectin (plant hemagglutinin) with blood group A specificity from Dolichos biflorus. Biochemistry 9:869.

Galbraith W, Goldstein IJ (1972). Lima bean (Phaseolus lunatus) lectin. Methods Enzymol 28(Part B):318

Hammarström S (1972). Snail (Helix pomatia) hemagglutinin. Methods Enzymol 28(Part B):368.

Imai Y, Oguchi T, Nakano T, Osawa T (1979). Separation of mouse T cell subsets by a fluorescent activated cell sorter using fluorescence-labeled peanut agglutinin. Immunol Commun 8:495.

Kaladas PM, Dabat EA, Kimura A, Ersson B (1981). The specificity of the combining site of the lectin from Vicia villosa seeds which reacts with cytotoxic T-lymphoblasts. Mol Immunol 18:969.

Kaufmann Y, Berke G (1981). Cell surface glycoproteins of cytotoxic T lymphocytes induced in vivo and in vitro. J Immunol 126:1443.

Kimura AK, Wigzell H (1978). Cell surface glycoproteins of murine cytotoxic T lymphocytes. I. T 145, a new cell surface glycoprotein selectively expressed on Ly $1^{-}2^{+}$ cytotoxic T lymphocytes. J Exp Med 147:1418.

Kimura AK, Wigzell H, Holmquist G, Ersson B, Carlsson P, (1979). Selective affinity fractionation of murine cytotoxic T lymphocytes (CTL). Unique lectin specific binding of the CTL associated surface glycoprotein, T 145.

Lang I, Banga JP, Varey AM, Gunn H, Cooke A, Roitt IM (1982). Isolation and functional characteristics of adherent phagocytic cells from mouse Peyer's patches. Immunology 46:769.

Lis H, Sela BA, Sachs L, Sharon N (1970). Specific inhibition by N-acetyl-D-galactosamine of the interaction between soybean agglutinin and animal cell surface. Biochim Biophys Acta 211:582.

MacDonald HR, Mach JP, Schreyer M, Zaech P, Cerottini JC (1981). Flow cytofluorometric analysis of the binding of Vicia villosa lectin to T lymphoblasts: lack of correlation with cytolytic function. J Immunol 126:883.

Nakano T, Imai Y, Naiki M, Osawa T (1980). Characterization of mouse helper and suppressor T cell subsets separated by lectins. J Immunol 125:1928.

Nakano T, Oguchi Y, Imai Y, Osawa T (1980). Induction and separation of mouse helper T cells by lectins. Immunology

40:217.
Pereira MEA, Kabat EA, Sharon N (1974). Immunochemical studies on the specificity of soybean agglutinin. Carbohydr Res 37:89.
Reisner Y, Linder-Iaraeli M, Sharon N (1976). Separation of mouse thymocytes into two subpopulations by the use of peanut agglutinin. Cell Immunol 25:129.

Index

PROGRESS IN CLINICAL AND BIOLOGICAL RESEARCH

Vol 1: **Erythrocyte Structure and Function,** George J. Brewer, *Editor*

Vol 2: **Preventability of Perinatal Injury,** Karlis Adamsons, Howard A. Fox, *Editors*

Vol 3: **Infections of the Fetus and the Newborn Infant,** Saul Krugman, Anne A. Gershon, *Editors*

Vol 4: **Conflicts in Childhood Cancer: An Evaluation of Current Management,** Lucius F. Sinks, John O. Godden, *Editors*

Vol 5: **Trace Components of Plasma: Isolation and Clinical Significance,** G.A. Jamieson, T.J. Greenwalt, *Editors*

Vol 6: **Prostatic Disease,** H. Marberger, H. Haschek, H.K.A. Schirmer, J.A.C. Colston, E. Witkin, *Editors*

Vol 7: **Blood Pressure, Edema and Proteinuria in Pregnancy,** Emanuel A. Friedman, *Editor*

Vol 8: **Cell Surface Receptors,** Garth L. Nicolson, Michael A. Raftery, Martin Rodbell, C. Fred Fox, *Editors*

Vol 9: **Membranes and Neoplasia: New Approaches and Strategies,** Vincent T. Marchesi, *Editor*

Vol 10: **Diabetes and Other Endocrine Disorders During Pregnancy and in the Newborn,** Maria I. New, Robert H. Fiser, *Editors*

Vol 11: **Clinical Uses of Frozen-Thawed Red Blood Cells,** John A. Griep, *Editor*

Vol 12: **Breast Cancer,** Albert C.W. Montague, Geary L. Stonesifer, Jr., Edward F. Lewison, *Editors*

Vol 13: **The Granulocyte: Function and Clinical Utilization,** Tibor J. Greenwalt, G.A. Jamieson, *Editors*

Vol 14: **Zinc Metabolism: Current Aspects in Health and Disease,** George J. Brewer, Ananda S. Prasad, *Editors*

Vol 15: **Cellular Neurobiology,** Zach Hall, Regis Kelly, C. Fred Fox, *Editors*

Vol 16: **HLA and Malignancy,** Gerald P. Murphy, *Editor*

Vol 17: **Cell Shape and Surface Architecture,** Jean Paul Revel, Ulf Henning, C. Fred Fox, *Editors*

Vol 18: **Tay-Sachs Disease: Screening and Prevention,** Michael M. Kaback, *Editor*

Vol 19: **Blood Substitutes and Plasma Expanders,** G.A. Jamieson, T.J. Greenwalt, *Editors*

Vol 20: **Erythrocyte Membranes: Recent Clinical and Experimental Advances,** Walter C. Kruckeberg, John W. Eaton, George J. Brewer, *Editors*

Vol 21: **The Red Cell,** George J. Brewer, *Editor*

Vol 22: **Molecular Aspects of Membrane Transport,** Dale Oxender, C. Fred Fox, *Editors*

Vol 23: **Cell Surface Carbohydrates and Biological Recognition,** Vincent T. Marchesi, Victor Ginsburg, Phillips W. Robbins, C. Fred Fox, *Editors*

Vol 24: **Twin Research, Proceedings of the Second International Congress on Twin Studies,** Walter E. Nance, *Editor* Published in 3 volumes: Part A: **Psychology and Methodology** Part B: **Biology and Epidemiology** Part C: **Clinical Studies**

Vol 25: **Recent Advances in Clinical Oncology,** Tapan A. Hazra, Michael C. Beachley, *Editors*

Vol 26: **Origin and Natural History of Cell Lines,** Claudio Barigozzi, *Editor*

Vol 27: **Membrane Mechanisms of Drugs of Abuse,** Charles W. Sharp, Leo G. Abood, *Editors*

Vol 28: **The Blood Platelet in Transfusion Therapy,** G.A. Jamieson, Tibor J. Greenwalt, *Editors*

Vol 29: **Biomedical Applications of the Horseshoe Crab (Limulidae),** Elias Cohen, *Editor-in-Chief*

Vol 30: **Normal and Abnormal Red Cell Membranes,** Samuel E. Lux, Vincent T. Marchesi, C. Fred Fox, *Editors*

Vol 66: **Control of Cellular Division and Development,** Dennis Cunningham, Eugene Goldwasser, James Watson, C. Fred Fox, *Editors.* Published in 2 volumes.

Vol 67: **Nutrition in the 1980s: Contraints on Our Knowledge,** Nancy Selvey, Philip L. White, *Editors*

Vol 68: **The Role of Peptides and Amino Acids as Neurotransmitters,** J. Barry Lombardini, Alexander D. Kenny, *Editors*

Vol 69: **Twin Research 3, Proceedings of the Third International Congress on Twin Studies,** Ligi Gedda, Paolo Parisi, Walter E. Nance, *Editors.* Published in 3 volumes: Part A: **Twin Biology and Multiple Pregnancy** Part B: **Intelligence, Personality, and Development** Part C: **Epidemiological and Clinical Studies**

Vol 70: **Reproductive Immunology,** Norbert Gleicher, *Editor*

Vol 71: **Psychopharmacology of Clonidine,** Harbans Lal, Stuart Fielding, *Editors*

Vol 72: **Hemophilia and Hemostasis,** Doris Ménaché, D. MacN. Surgenor, Harlan D. Anderson, *Editors*

Vol 73: **Membrane Biophysics: Structure and Function in Epithelia,** Mumtaz A. Dinno, Arthur B. Callahan, *Editors*

Vol 74: **Physiopathology of Endocrine Diseases and Mechanisms of Hormone Action,** Roberto J. Soto, Alejandro De Nicola, Jorge Blaquier, *Editors*

Vol 75: **The Prostatic Cell: Structure and Function,** Gerald P. Murphy, Avery A. Sandberg, James P. Karr, *Editors.* Published in 2 volumes: Part A: **Morphologic, Secretory, and Biochemical Aspects** Part B: **Prolactin, Carcinogenesis, and Clinical Aspects**

Vol 76: **Troubling Problems in Medical Ethics: The Third Volume in a Series on Ethics, Humanism, and Medicine,** Marc D. Basson, Rachel E. Lipson, Doreen L. Ganos, *Editors*

Vol 77: **Nutrition in Health and Disease and International Development: Symposia From the XII International Congress of Nutrition,** Alfred E. Harper, George K. Davis *Editors*

Vol 78: **Female Incontinence,** Norman R. Zinner, Arthur M. Sterling, *Editors*

Vol 79: **Proteins in the Nervous System: Structure and Function,** Bernard Haber, Jose Regino Perez-Polo, Joe Dan Coulter, *Editors*

Vol 80: **Mechanism and Control of Ciliary Movement,** Charles J. Brokaw, Pedro Verdugo, *Editors*

Vol 81: **Physiology and Biology of Horseshoe Crabs: Studies on Normal and Environmentally Stressed Animals,** Joseph Bonaventura, Celia Bonaventura, Shirley Tesh, *Editors*

Vol 82: **Clinical, Structural, and Biochemical Advances in Hereditary Eye Disorders,** Donna L. Daentl, *Editor*

Vol 83: **Issues in Cancer Screening and Communications,** Curtis Mettlin, Gerald P. Murphy, *Editors*

Vol 84: **Progress in Dermatoglyphic Research,** Christos S. Bartsocas, *Editor*

Vol 85: **Embryonic Development,** Max M. Burger, Rudolf Weber, *Editors.* Published in 2 volumes: Part A: **Genetic Aspects** Part B: **Cellular Aspects**

Vol 86: **The Interaction of Acoustical and Electromagnetic Fields With Biological Systems,** Shiro Takashima, Elliot Postow, *Editors*

Vol 87: **Physiopathology of Hypophysial Disturbances and Diseases of Reproduction,** Alejandro De Nicola, Jorge Blaquier, Roberto J. Soto, *Editors*

Vol 88: **Cytapheresis and Plasma Exchange: Clinical Indications,** W.R. Vogler, *Editor*

Vol 89: **Interaction of Platelets and Tumor Cells,** G.A. Jamieson, *Editor*

Vol 90: **Beta-Carbolines and Tetrahydroisoquinolines,** Floyd Bloom, Jack Barchas, Merton Sandler, Earl Usdin, *Organizers*

Vol 91: **Membranes in Growth and Development,** Joseph F. Hoffman, Gerhard H. Giebisch, Liana Bolis, *Editors*

Vol 92: **The Pineal and Its Hormones,** Russel J. Reiter, *Editor*

Vol 93: **Endotoxins and Their Detection With the Limulus Amebocyte Lysate Test,** Stanley W. Watson, Jack Levin, Thomas J. Novitsky, *Editors*

Vol 94: **Animal Models of Inherited Metabolic Diseases,** Robert J. Desnick, Donald F. Patterson, Dante G. Scarpelli, *Editors*

Vol 95: **Gaucher Disease: A Century of Delineation and Research,** Robert J. Desnick, Shimon Gatt, Gregory A. Grabowski, *Editors*

Vol 96: **Mechanisms of Speciation,** Claudio Barigozzi, *Editor*